Understanding COVID-19 Vaccination Arguments

G. G. Bolich, Ph.D.

EVS PRESS
2021

EVS Press
Spokane, Washington
©2021 G. G. Bolich

All rights reserved. No part of this book may be used or reproduced in any manner whatsoever without permission, except in the matter of brief quotations employed in reviews or critical works. For permissions, please contact the author at gregbolich@gmail.com.

For

Brook.

Publisher Cataloging-in-Publication Data

Understanding COVID-19 vaccination arguments.
Bolich, G. G. (1953-)
First paperback edition.
ISBN 13: 978-1-365-82231-5
1. Vaccines 2. Public Health
I. Title
QR189.5.B65 2021
614.47.B689—dc23

Printed in the United States of America
1st edition

Table of Contents

Preface	1
Introduction: Plan of the Book	3
1. How to Know (and Assess an Argument)	5
2. Meeting the Enemy	11
3. The Heart of the Debate	17
4. The Pro-Vaccine Arguments	27
5. The Anti-Vaccine Arguments	37
6. The Pro-Vaccine Rebuttal	55
7. The Anti-Vaccine Positive Position	91
8. An Intersection at Ethics?	101
9. The Bottom Line	109
10. Free Advice (Worth the Price!)	119
Appendix: Signees of a Joint Statement	125
Bibliography	127

About the Author

My name is Greg Bolich, and like any author I would like your trust. Unlike some authors I hope to *earn* your trust. So in that spirit let me tell you a little about myself.

Raised in very conservative Christianity, I was educated first at an Evangelical college, then a conservative seminary, then a Jesuit university, and finally a private university. After achieving a B.A. with a double major in philosophy and religion, I earned five graduate degrees, including two doctorates. Among the important skills I bring to this work from my education is a strong background in *epistemology*—the study of knowing and knowledge. My second Ph.D. focused on biological psychology (sometimes called "medical psychology"), including advanced work in neuropsychology and culminating in a dissertation on the neurobiology of psychological boundaries.

I became, I imagine it could be said, a 'professional scholar,' trained in research, critical thinking, and objective reporting. I have authored numerous published articles and books in several fields. As a university teacher I taught graduate level courses in psychopharmacology and various courses related to research.

That said, I am neither a medical doctor nor an immunologist. I am, however, well-positioned to read and understand them and report their work. My educational background and the demands of subsequent professional obligations resulted in a thorough acquaintance with medical history, terminology, and research. Some of this acquired learning has been previously published in various of my writings. But please understand that I am not asking for your trust based on my being some kind of authority. I want your trust based on the reasoning and the evidence provided herein—nothing else.

Again, unlike some writers who reach beyond their level of competence, I have stayed carefully within mine and drawn heavily on others whose expertise exceeds my own in various matters relevant to understanding the subject at hand. But I wish to repeat that the goal is not to appeal to authorities except in acknowledging their accurate handling of facts accessible to all of us. There has been entirely too much of people pitting so-called 'authorities' against one another. Usually all that is meant is naming people with degrees who say what a person wants to hear. Real authority resides in truthful facts, which don't much care whether we agree with them or not.

I do hope the diverse personal, educational and professional background I bring to the task of this book has uniquely positioned me to understand both sides and to present a clear, cogent, and fair appraisal of the arguments about COVID-19 vaccination. That is not to say, however, that I find all the arguments on both sides to be equal in weight when measured. Like anyone else, I want to know what is true, and make a wise decision based on that knowledge. In that spirit, 'Come, let us reason together.'

Preface

This is not a book I wanted to write.

Like so many Americans, my family has been deeply divided by the culture conflicts in our country. These conflicts have intruded into many areas of life, and in an unprecedented way into debates over public health. These debates are played out not only on a national stage, but on very private family ones as well.

In my own extended family are several health care professionals who have worked in hospital settings during the COVID-19 pandemic. There are other family members who have been sickened by the virus. Some in my family strongly support vaccination, while others just as strongly oppose it. Some are what the media have labeled as "vaccine hesitant," meaning they are uncertain as to whether to be vaccinated or not. In short, my family mirrors the larger American family.

With the COVID-19 pandemic and the horrific loss of life it has brought, all Americans have found themselves having to wrestle with important questions about how best to care for their own health and that of loved ones for whom they are responsible. Because the stakes are so high—quite literally, life-or-death—emotions often run hot when discussions are held about topics like vaccination against the virus and its variants.

Unfortunately, intense feeling coupled with an American preference for quick and decisive choices can produce judgments that are strongly held but only shallowly considered—on either side of the debate. Discussions often tend to shed more heat than light.

One advantage of a book is that it provides *time*. Reading is done privately, requires sustained attention, and offers the opportunity to slow down and reflect on matters in a way seldom, if ever, provided by face-to-face discussion or by chatting through texting.

I am certainly not alone in being someone who has been involved in many discussions about COVID-19. The pandemic's affect on American life is multi-dimensional and touches us all. Accordingly, in my family we have talked about things like social distancing, wearing masks, government mandates, and the safety of our family members, especially the most vulnerable—our children and grandparents, as well as those with underlying health conditions.

There is no lack of things I might choose to write upon with respect to COVID-19. And, no matter what I might choose, some in my family will end up agreeing with me and others disagreeing. What I regret most is realizing some of those I love will choose to not consider anything written here, both because they have made up their minds and are closed to further consideration, and because they are certain I cannot say anything more in writing than I have said already in firsthand discussions.

Yet others in my family have resolved to be open-minded. I have been asked by them to gather together and present in writing information about one particular COVID-19 matter: *vaccination*. The matter sits at the very center of the most intense scrutiny by Americans and is the focus of some of loudest arguing in my family. In this small book I have attempted to as simply as possible discuss the vaccination debate, laying out the arguments presented by those for and against it, and doing so fairly and in a manner that facilitates reconsideration where needed and perhaps more thoughtful appraisal.

Shall we begin?

Introduction
Plan of the Book

It seems only fair to spend a couple pages on explaining my plan for this book. That may help you decide how to approach reading it and using it.

Intended Audience

I am presuming open-mindedness to further consideration of the subject at hand. That presumption seems warranted, at least for most people; only a minority say they have reached a point where nothing can change their minds. Americans are still fluid in their thinking about many issues related to COVID-19. The Axios-Ipsos Coronavirus Index[1] released August 31, 2021 indicated continuing changes in the United States both in concern over the pandemic and in thinking about the vaccine.[2] This is a hopeful sign.

Those who have made up their minds and will not change are unlikely to bother spending much time here if they determine I am not saying what they want to hear. Such people pretty uniformly give lip service to 'research' and instead focus on reinforcing self-indoctrination by only listening to those who echo their own beliefs. Since it is impossible to have a genuine dialog of substance with them, I prefer to follow Jesus' words (Matthew 7:6). Neither am I interested in 'preaching to the choir' and fueling zealots who want merely to have more ammunition to fire at 'the enemy.'

There are no enemies here save ignorance and deception. Those who have allied themselves with such forces will in due time reap the consequences.

My intended audience is those who genuinely want to *understand* the issues involved in the vaccination arguments. These are many, sometimes complex, and often obscured by parties on both sides, whether through misunderstanding or a desire to 'win' at any cost. My hope is to cut through the confusion and make things a bit plainer to facilitate clearer, better thinking.

I am appealing to those of open mind to keep learning and make informed decisions, and to accept that all decisions should remain open to the possibility of reversal in the interest of truth.

I respect open-minded people of good will on every side no matter how wrong I may judge their thinking and conclusions at the moment. The only people I really fear are those so confident of possessing the truth they feel they have no need to listen anymore. Too many see truth as a weapon to wield against others. Far better we see truth as a liberating light. In the interest of finding light rather than generating heat, I have a plan.

---

[1] Axios is a fact-based journalism site (https://www.axios.com/about/); Isos is a multinational market and research firm (https://www.ipsos.com/en). They have partnered to produce a weekly Coronavirus index based on polling data (https://www.openicpsr.org/openicpsr/project/129181/version/V1/view;jsessionid=80053828D4C34C509E592D874BDD61CD).

[2] Ipsos, "The Wall of Vaccine Opposition."

How to Read this Book

One need not read this book in a linear fashion to profit from it. Each chapter stands on its own and presents a manageable amount of material on a particular part of the whole. I imagine some will be impatient to get immediately to the arguments themselves, and that is okay. Still, the book is designed to present what I hope is a sensible progression. After considering differences in how people claim to know truth, I turn briefly to some guiding points about understanding how arguments are made and can be better assessed (chapter 1). Then we meet the enemy—the virus that leads to the disease COVID-19 (chapter 2). Next I try to highlight what are the key concerns, first, of pro-vaccine people, and then of anti-vaccine adherents (chapter 3). The arguments themselves are spread across four chapters, starting with the pro-vaccine position (chapter 4), then anti-vaccine arguments against it (chapter 5), the pro-vaccine effort to rebut those arguments (chapter 6), and finally the positive case of what anti-vaxxers affirm (chapter 7). Next I propose there may be some fruitful common ground in ethics (chapter 8). Following that I render my own assessment of the strength of the arguments (chapter 9), and offer some free advice to all parties (chapter 10).

I have tried to be transparent in the presentation of materials. Rather than do what tends to be done in ordinary conversation—simply swap opinions—I have tried to be objective and fair. That means I have drawn upon a *very* diverse set of sources and documented what they are. I have intentionally utilized sources that can be tracked down on the internet so that anyone desiring to verify my reporting can check for themselves to see if I have been fair. Due diligence should be done in vetting any source used.

I hope readers will take the time to track down materials. Even more I hope readers will not confine themselves to only listening to those who confirm what they already believe. Although I grew up in a very conservative environment, I was still encouraged to listen respectfully and openly to others. It is okay to change one's mind when there is sufficient cause to do so.

I know, as we all do, that *not all arguments are equal*. In fact, relatively rarely does it prove to be the case that both sides in a debate are equally fair and accurate in their handling of facts, equally plausible in their reasoning, and equally interested in promoting the common good. It is a mistake to represent opposing sides as equal if, in fact, they are not. Nevertheless, even poor arguments deserve fair and undistorted presentation, by their advocates.

It's also a mistake to start from the presumption that one's own inclination is superior and to judge opposition as misguided or deficient without even considering it. A judgment of arguments should *follow* examination of them, not precede it. Such judgment is most credible when well-informed about each of the opposing sides and also well-prepared by an understanding of how people decide they know what is true and understanding how arguments are best assessed, which is where we shall begin.

Chapter 1
How to Know
(and Assess an Argument)

Sometimes it seems like everyone not only has an opinion on COVID-19 vaccination, but they are sure they are right because, after all, who holds an opinion he or she knows is *wrong*?

It is easy to presume one's own opinion is right and that only those who agree with us are also right. When it comes to high-stakes arguments, it is also easy to think that others are not only wrong, but perhaps willfully and evilly wrong. Sometimes that is the case but since we have the luxury of some private reading time, let's agree to suspend such initial judgments. In fact, let's start a little distance away from the topic itself and give ourselves a few minutes to start really basic. (If it seems too basic, skim this and skip ahead.)

How Do We Know?

Let's ask, 'How does anyone *know* something to be true?' It's a question that has occupied philosophers for thousands of years, and I won't pretend we'll settle it here. But it is a good place to start because many arguments come down to differences in how this question is answered—and it is crucial in the vaccine debate.

In broad strokes, most people who think something can be known appeal to one or more of the following sources:

❖ authority;
❖ evidence from our senses;
❖ reason; and/or
❖ intuition.

Across our lives we justify conclusions, decisions, and actions based on first one source, then another, and quite often more than one used together.

Authority is a common and broad source. When we have car trouble and decide going to a mechanic seems a smarter choice than calling our banker we are appealing to the *authority of expert knowledge*. We credit learning to develop professional skills and then experience practicing them as together conferring a degree of knowledge we do not ourselves possess but can profit from. Polls consistently show that the stance toward vaccination that Americans have is positively correlated with their trust or lack of trust in authorities such as physicians, scientists, politicians, and media people.[3]

The authority of expert knowledge is not the same as the authority of *person*. The latter rests on position, esteem, or credentials, but not expert knowledge as

---

[3] For details, see nonpartisan poll reports such as those from the Pew Research Center. Good places to start include Dean, Parker, and Gramlich, "A Year of U.S. Public Opinion," Funk and Tyson, "Intent," and/or Funk and Tyson, "Growing Share."

such. Appeals to this kind of authority should not be confused with the preceding sort. Simply because, for example, someone has an advanced degree does not in itself guarantee an expertise of knowledge. Authority of knowledge requires demonstration of knowledge.

The source called *revelation* is one form of authority and regarded by many Americans as the highest form. Since the preeminent religion in the United States is Christianity, appeals to the Bible as the source of knowledge most to be trusted and turned to is not uncommon. Some Americans prefer to rely on biblical guidance—often by way of some trusted Christian figure—rather than other sources, even if they also consider those other sources.

Evidence from our senses is basic to moment-by-moment living. If we place a hand on a hot stove we don't wait for an authority figure to teach us what to do next. The unpleasant experience provided by our senses is a sufficient source. The accumulation of such evidence underlies science.

Reason is the ability to think along logical lines. It makes possible the organization of evidence from the senses to produce testable theories. Thus it is the partner of sense evidence in science. For those who rely on God's revelation there is a strong sense that reason is God's gift to assist in understanding and applying the divine will. Thus reason is held by all to be important.

Intuition commonly refers to what some people call 'gut instinct,' or a 'sudden hunch,' or 'insight.' Formally it refers to an immediate and direct awareness of something apart from any other influence, marked by an acute sense of certainty. It can be harder to describe than the other ways of knowing, and harder to explain. It often lies behind people saying, "I just *know*."

Each of these sources persist as ways people use to know things because no single source is always used; ordinary experience demonstrates the value of using more than just one.[4]

Skepticism

Of course, some people are *skeptics*. They argue nothing can be 'known' in the sense of achieving certainty. Some decide to suspend judgment on everything. That may work for many matters, but the practice of living often requires making a decision whether we know it is the right one or not, and whether we have certainty. The question of COVID-19 vaccination is one such matter. If one chooses to just wait-and-see, and justify it with a skeptical stance, that is, as they say, 'voting with one's feet.' Inaction has its own consequences. In that sense, suspending judgment might be dangerous.

Others are skeptical in a different sense. Recognizing how difficult it can be to achieve knowledge about something, they decide to mistrust all claims, weigh them all equally (i.e., as "unproven"), and thus simply decide based on whatever seems best in the moment. This highly flexible way of thinking and deciding

---

[4] The subject of knowledge and how we might know best is a weighty one and well worth further time and effort. It has occiped me across several volumes. One of these, *Knowledge: An Illustrated History*, provides a basic overview of the major positions and thinkers about this matter.

often works well enough, especially where the personal stakes are low. In ambiguous situations it can lead to deferring action, which might be wise. However, in high-stakes situations where delay heightens risk, or in cases where some claims are more probably true, this kind of skepticism can produce impulsive, foolish decisions.[5]

Knowledge, Certainty, and Belief

Consider the phrase above, "where some claims are more probably true." This raises a concern skeptics sometimes highlight: should we even talk about knowledge as a matter of *certainty*? Science seems to suggest we never have more than mere probability. On the other hand, divine revelation promises certainty is possible and should guide our every decision and action. Such contrasting positions make it inevitable that for some people the matter of deciding about the COVID-19 vaccination comes down to *faith versus science*.

Yet most people cling to the hope that faith and science can be partners rather than opponents. Moreover, most people, even while yearning for certainty, are content in many things to rely on probability to establish knowledge of something. Thus most people are sympathetic to Aristotle's basic notion that there are *degrees* of knowing ranging from probable to certain.

Aristotle started from a conviction that "All people by nature stretch themselves toward knowledge."[6] But he insisted that our stretching is often limited. Knowledge that does not reach the level of certainty can still be productive and practical. He also left a place for opinions and beliefs—at least when they are reasoned ones. For example, he noted that most of us distinguish between the opinions held by uninformed people and those espoused by the most learned or best informed. Thus he formulated a hierarchy about knowing and knowledge that resists skepticism and acknowledges that belief and probability both have roles in knowing.

Aristotle also is famous for laying down ideas about arguments that have guided thinking about them ever since. We don't need to go into those or adopt an Aristotelian view on either knowledge or arguments, but we do need to take a moment to consider the basic question of how we should assess arguments.

Understanding Arguments

To reiterate, the truth is that *not all arguments are equal*. Sometimes the inequality of competing arguments is obvious. For example, the vast majority of people judge that arguments for a 'flat earth' are far less compelling than those against such a view.

Assessing any argument can be challenging, even though most of assume we do well at the task. Still, no matter how we judge our skill, perhaps we might agree on employing a SIMPLE Method.[7] Here are its elements:

---

[5] Skepticism has a long history and more than one form. For more, see Bolich, *Knowledge: An Illustrated History*, chapter 7, or Bolich, *Knowledge and Belief*, chapters 4 and 10.
[6] Aristotle, *Metaphysics* 980.a21 [I.1]. On his position, see Bolich, *Knowledge and Belief*, 101–31.
[7] The SIMPLE Method is explained in Bolich, *Understanding Arguments*.

- ❖ **S**ources of information vary in credibility.
- ❖ **I**ntention signals the aim of an argument.
- ❖ **M**essages provide both information and meaning.
- ❖ **P**ersuasiveness reflects a style of arguing.
- ❖ **L**ogic points to the quality of thinking.
- ❖ **E**motions provoked by an argument offer valuable feedback.

These six elements can be considered in any order one likes. In fact, any given argument might prompt one particular element to be considered first or judged most important because it is either strong, or weak, or even absent.

The *sources* used by an argument affect credibility. Generally speaking, a person who uses him- or herself as a sufficient source, disdaining all others, is likely to be less credible (even if an expert) than someone who appeals to a variety of credible sources. This is one reason students are taught to investigate and use multiple sources when they do research. Even those who rely on divine revelation look to more than one text in the Bible to reach a decision on important matters.

Several factors are important to assessing source credibility. An obvious one is the source's credentials. A mechanic talking about car engine trouble is more credible than a tax accountant doing the same. We credit learning, typically demonstrated by educational achievement and professional experience.

Another factor can be the number of concurring sources on a point. For instance, peer review—the assessment of an argument or study by other experts—is often used as a basic measure of credibility. Or in reading the Bible finding Old and New Testaments saying the same thing is strongly confirming of God's voice in both.

However, it is important to remember that this can be easily manipulated. Websites, for example, often repeat information from a common source so that what seems like a hundred points of agreement is really only one claim endlessly repeated. Ancient philosophers recognized this problem and pointed out that a 100 people saying the same thing is still just *one* point of view. And, sometimes the one voice against the hundred is the right one. Still, most of us would say that generally speaking it seems more sensible to side with 100 experts than 1 nonexpert when there is a disagreement.

Similarly, some people pick and choose texts to make the Bible say what they want it to say. Skilled interpreters of revelation remind us that *context* is critical to keeping information—whether from the Bible or anywhere else—properly framed. In arguments context always matters.

Four closely related factors are especially critical in assessing sources. First, does the source make vague or general claims, or offer specific and reasonable claims? Second, does the source show any openness to alternative explanations, to weaknesses in its own position, and to the possibility of new evidence? Third, does the source offer ways it can be tested, for example by setting down a clear demonstration of how its conclusions were reached so that anyone can check out what is being claimed? Fourth, are the methods used by the source ones that have been found reliable and valid?

If a source is specific and fact-based, open to new evidence while aware of its limitations, shows us how the conclusions were reached, and uses time-tested methods endorsed by fruitful results, then it makes a good case for its credibility. A source isn't credible simply because it claims to be. Any source that isn't open to question or challenge is suspect.

The *intention* of the arguer matters. Some arguments have a profit motive. Everything presented is intended to help make a sale or otherwise generate something personally valuable for the arguer (e.g., money, public esteem, or winning an election). Other arguments intend to set out, as best they can, what is actually the case of a matter. Many arguments aim at persuading that the contention being made is endorsed by most people—or at least the most well-informed ones. Others intend to confirm to the like-minded in their audience that they are right; their shared presumptions are all that matter.

It is easy enough to see that different intentions can lead to different sources being used and produce different *messages*. This third element is a compound one made up of both information content (the argument's support) and the meaning the argument derives from that content (the argument's claim).

This means we need to determine what exactly makes up the information content and what exactly is the meaning being claimed. Then we need to determine whether the meaning depends on the information content or if the information is just window dressing. A little sleight of hand can make a message superficially *seem* to rely on information content, but closer looking shows it doesn't rely on that information at all. Thus the final step is forming a judgment that asks and answers one question: does the argument's information content actually support the meaning claim made?

Persuasiveness depends in part on all of the other elements. Credible sources, clear intention, and coordinated messages all foster persuasiveness. So, too, can logic and emotional appeals. But arguments can be persuasive for all kinds of reasons. Some people are persuaded, for example, simply because the presenter of the argument is attractive. So this element is mostly about figuring out *why* an argument is or is not persuasive to us. Doing so can help us become wiser about *how* we let ourselves be persuaded—and that can help us avoid decisions we might later regret, whether buying a car or deciding for or against a vaccine.

Logic is something we all employ. I don't mean just formal logic, like that found in syllogisms (e.g., "All men are mortal. Socrates is a man. Therefore, Socrates is mortal."). Formal logic is good, but most of the time we use informal logic, as in what is termed 'presumptive reasoning.' That means we start from some presumption; these don't actually depend on careful reasoning but rely on the listener accepting some initial claim as plausible and, especially, desirable. Most arguments one encounters use informal logic.

Put simply, these informal logic arguments are the kind neighbors and family members usually make. For example, I might make a claim that my case is plausible and desirable because of its likeness to an analogous case. This is the sort of thing done when lawyers appeal to precedence to help decide a case. Or I might argue that my case is better because my authority is higher than my opponent's. Or I might argue that because my audience has already committed to an idea, and mine is just an extension of that, they should agree with my idea—a classic style of argument used in sales pitches.

There are all kinds of such arguments based on one form or another of informal logic. The key is to remember that *none of them actually depend on reason or evidence*. They stand or fall on whether the presumption being made is accepted. Thus, in an argument appealing to an authority the presumption is that the authority knows more and better than we do; the actual case (good or bad) made by the authority is secondary. This way of arguing is not inherently awful, and can lead to defensible conclusions, but it does have its hazards. We may think we are being persuaded by one thing (e.g., reason and evidence) when we are actually being convinced by something else (the presumption).

Finally, *emotions* are important to consider. Some people are dismissive of them; others over-value them. It is valuable to regard emotions as an important source of information, but neither the *only* important source nor always the *most important* one. Feelings generated by an argument tell us more about our response than the actual merits of the argument. If we can take a moment and make ourselves step back from a strong emotional response we might to our surprise discover that the argument has more merit—or less—than we first felt.

Conclusion

These few pages are intended to help in evaluating arguments about COVID-19 vaccination. In brief, start by considering how both sides determine one might know the truth of the matter. Widely different ways of trying to know something can make understanding each other difficult for the parties involved. Then try to apply the elements of the SIMPLE Method to understanding the argument and assessing its worth.

Using the material in this chapter can help us better understand arguments about COVID-19 vaccination and to making (or reconsidering) personal decisions about it. But before we turn to the arguments themselves, we need to meet the enemy about whom people are arguing as how to best fight it. That enemy is a virus, and while discussing viruses quickly engages us in vocabulary not generally a part of everyday talking, it is well worth taking time to slow down, plow through, and gain at least the gist of what this enemy is all about and how it wages war against us.

Chapter 2
Meeting the Enemy

One of the colorful descriptors often encountered in discussions of the pandemic is that humanity is at "war" with COVID-19. The "enemy" is a disease that has killed millions, severely sickened millions more, and shows no signs of surrendering any time soon. To understand this enemy—this disease—we must spend some brief time looking at the virus that causes it.

SARS-CoV-2

Most people today are aware of the difference between *bacteria*, single-cell organisms, and much smaller *viruses*, which are comprised of small snippets of genetic code. Viruses are puzzling with respect to a very basic question: are they alive? The question draws both 'yes' and 'no' answers from scientists. We need not answer the question here, but it is one important difference between bacteria—clearly living organisms—and viruses. Another difference, and important to us here, is that bacterial diseases are commonly susceptible to treatment by antibiotics, but viruses are not. So when a virus such as the one that causes COVID-19 occurs, antibiotic remedies to cure it are not available.

The virus that produces the disease COVID-19 is labeled SARS-CoV-2. The label is a short form for the name 'Severe Acute Respiratory Syndrome (SARS) Coronavirus (CoV) 2.' The first part of the name describes the *effect* of the virus: it causes a characteristic group of symptoms ('syndrome') associated with the human body's system for breathing ('respiratory') that occurs suddenly ('acute') and with great intensity ('severe'). The next part of the name describes the *biological viral family* to which the virus belongs, called coronavirus because under an electron microscope its surface shows distinctive crown-like features that produce a visual effect like looking at the sun's corona. The number '2' distinguishes it from the previously discovered SARS-CoV virus.

The Coronavirus Family

SARS-CoV-2 belongs to a family of viruses called the 'Coronaviruses.' They infect human beings and many other animals. Seven coronaviruses are known to infect humans. Four of these generally produce in people what we term the common cold; the other three create more serious disease. The seven are[8]:

Less Serious	*More Serious*
HCoV 229E; HCoV NL63; HCoV HKU1; HCoV OC43— all causing cold symptoms	MERS-CoV (causing Middle East Respiratory Syndrome); SARS-CoV (causing Severe Acute Respiratory Syndrome); SARS-CoV-2 (causing COVID-19)

---

[8] Hasöksüz, Kiliç, and Saraç, "Coronaviruses and SARS-CoV-2," 549. Cf. Hu et al. "Characteristics of SARS-CoV-2 and COVID-19," 142–43.

Coronaviruses have been known since the start of the 1930s, though they were not formally called 'coronaviruses' until the late 1960s, after the discovery of the first coronaviruses to infect humans (HCoV 229E and HCoV OC43). While these viruses received some attention—including the entire coronavirus genome sequencing in 1987—it was the emergence of SARS-CoV and the epidemic it spawned at the start of the 21st century that drew sharp attention and some urgency because of its high fatality rate (10%). It was this new diligence that identified two more human coronaviruses (HCoV NL63 and HCoV NL63). When MERS-CoV emerged in 2012, with a dramatic mortality rate of more than 1-in-3 (36%), the coronavirus family became even more a matter of serious worldwide scientific scrutiny.[9]

SARS-CoV-2 and Basic Genetics

Like any virus, SARS-CoV-2 consists of a relatively short sequence of genetic material. Although somewhat longer in sequence than many viruses, it is still only about 29,800 nucleotides (each an organic molecule made of adenine, guanine, uracil, or cytosine) that encode 27 proteins. (Don't get bogged down in the unfamiliar terms; focus on the fact that *SARS-CoV-2 is a very small package with limited abilities*.) Technically, SARS-CoV-2 is an "enveloped positive-sense single-stranded RNA" virus (+ssRNA). Without going into detail, this refers to how the virus stores its genomic information (and we will touch on that below).

Now we need to review some basic genetics about human beings. We all have learned that what we are depends on our DNA (deoxyribonucleic acid). Think of DNA as providing instructions that our body's cells will then carry out. These instructions are coded by genetic letters (A, G, C, T)—the alphabet of genetics. This alphabet has to be 'transcribed'—put into a workable form so the instructions of DNA can be carried out. To get from the instructions to the actual work done by the cell, DNA needs a helping partner.

Portions of DNA, like letters used to form words, are transcribed by RNA (ribonucleic acid) as what we call 'genes.' Most genes contain the DNA information required to make 'proteins'—complex molecules that carry out the work of the cell. We might say that RNA 'reads' DNA material in genes and converts the coded sequences into 'work orders' carried out by proteins.

RNA comes in three different types. Messenger RNA (mRNA) makes one specific 'work order' carried to a cell's ribosomes, its manufacturing apparatus outside the nucleus. Each mRNA acts like a Snapchat message—it has a limited, specific message of very short duration. Transfer RNA (tRNA) carries to the ribosomes material needed for protein manufacture. Ribsomal RNA (rRNA) is the predominant stuff of which ribosomes are made.

The above are basic actors in the genetic work chain in human cells. We might liken this to an architect with blueprints (DNA) using a general contractor (RNA) to manage workers (proteins). As we saw a moment ago,

---

[9] See V'kovski et al., "Coronavirus Biology," 157, Box 1.

SARS-CoV-2 is a *RNA virus*. Put metaphorically, it is a general contractor with a limited skill set, but what it does, it does very well. It hijacks the cell it enters and changes the work orders in its own favor.

But to understand that process we will profit from looking more closely at SARS-CoV-2. The Center for Disease Control (CDC) has prepared an illustration of the virus highlighting certain structural proteins: [10]

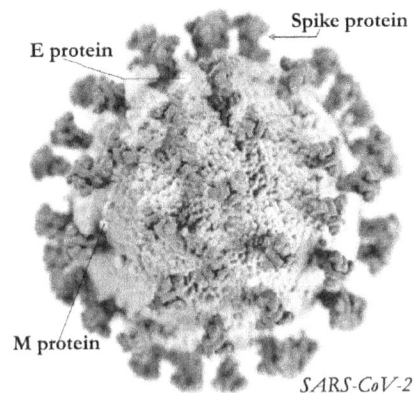

SARS-CoV-2

The so-called 'S protein' is the distinctive spike which dots the virus' surface and identifies it as a coronavirus. It is used to attach to a host cell.

The E (Envelope) protein is named for its function of creating an outer wrapping (the 'envelope') from the infected cell.

The M (Membrane) glycoprotein is the most abundant protein and works with E; it defines the shape of the envelope formed.

Coronaviruses also have another prominent structural protein (not labeled above). The N (Nucleocapsid) protein is essential for the virus' replication and a part of the viral life cycle known as 'genome packaging,' whereby the replicated genome of the virus is placed into a protective envelope.[11]

How SARS-CoV-2 Infects Human Host Cells

SARS-CoV-2, as a 'positive-sense' RNA, can act as mRNA, meaning it can send the very specific message to an infected cell to make the proteins the virus desires (i.e., virus proteins) rather than what the host's DNA intends. Using the language of our above metaphor, it subverts the cell's intended work orders. Instead of carrying out its normal functions the host cell is turned into a factory for replicating and sending out the virus.[12]

SARS-CoV-2 uses a two-step process to enter a human host cell. First, the virus uses its S (Spike) proteins to attach to a specific place on a cell. The spike proteins have three points at their top that bind to surface points on a cell called 'receptors.' The S protein chemically recognizes a very specific kind of receptor, labeled hACE2 (human angiotensin-converting enzyme 2). The portion of the virus Spike protein called the 'receptor-binding domain' (RBD) rises to a

---

[10] CDC, Public Health Image Library (PHIL) #23312. The image has been slightly adapted by including identification of some features and changed from color to grayscale. For brief descriptions of all four structural proteins, see Schoeman and Fielding, "Coronavirus Envelope Protein," 2. For detailed descripton of the virus' transcriptome, see Kim et al., "The Architecture of SARS-CoV-2 Transcriptome."
[11] Cubuk et al. "The SARS-CoV-2 Nucleocapsid Protein."
[12] For the science of this process, see Sanyal, "How SARS-CoV-2 (COVID-19) Spreads," and Mattheson and Lerner, "How Does SARS-CoV-2 Cause COVID-19?"

standing-up position to bind to a pocket on the hACE2 receptor. Second, the virus also employs cell fusion. The viral envelope (see proteins E and M, above) fuses with the host cell's membranes and thus transports the virus genome into the host cell's cytoplasm (the fluid inside the cell membranes).[13] To return to our metaphor, the virus finds a door into the worksite (step 1: attachment to a receptor), then sneaks in (step 2: fusion).

Part of what makes SARS-CoV-2 so successful is that it employs a process at the hACE2 site that helps it both bind to the host cell *and* avoid detection by the host body's immune system. To bind successfully the S protein's RBD has to be in a standing upright position. But that position can flag it to the immune system. So the S protein varies between this standing-up position for binding, and then lying down to evade detection. Combined with its manner of cell fusion to inject its own genome, SARS-CoV-2 is able to gain rapid entry, reproduction, and spread while evading detection.[14]

Understanding this process has practical consequences. *Where* the virus enters the host to do its work helps explain *why* certain parts of the body are severely affected. The hACE2 receptor plays a key role in regulating blood pressure and it is found in many places in the human body, most notably the epithelial cells of the respiratory system. But hACE2 sites also are found in the cardiovascular system, gastrointestinal tract, and in the kidneys, liver, and nervous system. Impairment of hACE2 in the cardiovascular system, for example, can lead to severe cardiac dysfunction. Similar ill effects can be found in the other places jut mentioned when hACE2 function is compromised.[15]

To switch to the metaphor this chapter opened with, SARS-CoV-2 is an enemy expert at infiltration and rapid deployment. It quickly spreads its troops before an alarm is spread. By the time the host country—the human body—begins to muster its own troops (the immune system), significant damage may have occurred. What that damage might look like is better understood when the above description of how the virus infects its host is comprehended.

SARS-CoV-2 and COVID-19

SARS-CoV-2 causes the disease named COVID-19, first noticed in December, 2019. It was referred to at first as a 'novel' coronavirus, because it was *new*, having not been previously known. Although it was circulating before it became publicly identified, once it was named as SARS-CoV-2 (February 11, 2020) the disease whose spread it prompted was quickly named COVID-19 by the World Health Organization (WHO). The disease spread from Wuhan, the capital of Hubei province in central China. The virus was quickly and

---

[13] On the general topic of entry into a host, see Shang et al., "Cell Entry Mechanisms." On hACE2, see Salamanna et al, "Body Localization of ACE-2." On the fusion process, see Millett and Whittaker, "Host Cell Proteases."
[14] See Shang et al., "Cell Entry Mechanisms."
[15] Salamanna et al., "Body Localization of ACE-2," 3. Also see Figure 1 (p. 4). See the entire article for specification of effects in different organs.

extensively studied, with the first genome sequence published January 10, 2020 and more complete genome sequences two days later.[16]

COVID-19 remains most widely known as a respiratory illness with symptoms like those associated with pneumonia. These include:
- cough (including coughing up blood);
- fatigue;
- fever;
- headache;
- myalgia (muscle pain);
- phlegm coughed up from the lungs (sputum production); and,
- shortness of breath.

The above list reflects the invasion of epithelial cells in the respiratory tract, which then leads to virus replication and migration downward to the airways, where it enters alveolar epitheal cells in the lungs. The virus' invasion of an ill-prepared immune system can lead to such a strong (one might metaphorically say 'panicked') response that a condition known as 'cytokine storm syndrome' occurs—a life-threatening inflammatory response that kills tissues the immune system ordinarily protects. The result is the severe acute respiratory syndrome (SARS) that is the first part of the virus' name.[17]

The above list poses problems enough. However, it quickly became apparent that COVID-19 also is associated with a disturbingly wide-range of other symptoms, across a range of organs, including[18]:

Organs affected	*Symptoms (arranged alphabetically)*
Brain	brain inflammation, confusion, seizures, and strokes
Eyes	conjunctivitis, inflammation of the membrane that lines the front of the eye, and inner eyelid
Heart and blood vessels	arrhythmias, blood clots, cardiac inflammation and injury, chest distress/pain, elevations of cardiac injury biomarkers, heart failure, lymphopenia (an abnormally low number of lymphocytes (a type of white blood cell)), and palpitations (palmus)
Intestine and stomach	abdominal pain, anorexia, diarrhea, nausea, and vomiting
Kidney	abnormal creatinine level, protein and blood in urine
Liver	abnormal enzymes levels
Nose	partial or full loss of smell (anosmia)
Pancreas	pancreatitis

These are all places where hACE2 receptors on cells are found.

---

[16] Hu et al., "Characteristics of SARS-CoV-2 and COVID-19," 141–42.
[17] Hu et al., "Characteristics of SARS-CoV-2 and COVID-19," 147.
[18] Both the pneumonia associated symptoms and symptoms of other organs are adapted from Hu et al., "Characteristics of SARS-CoV-2 and COVID-19," 148, and Salamanna et al., "Body Localization of ACE-2," 1.

COVID-19 also soon became associated with a distressing condition known as *long COVID*, in which patients continue to be plagued by symptoms months after infection. Moreover, this condition has been found even in young, low risk patients. For example, a late 2020 report in England found that almost 70% of 201 patients in this group continued to experience impairments in one or more organs four months after initial symptoms.[19]

As noted earlier, as an enemy SARS-CoV-2 is especially worrisome because of its high success in evading detection while rapidly spreading. But added to that reality is that it is highly transmissible from one person to another because the viral load in the host is so high even before the person feels sick. The 'serial interval' (i.e., the time between symptom onset in the first infected person and the symptom onset in the next infected person) median is about four days. That is actually a shorter time than the estimated incubation period, which means *transmission of the virus to others is likely before the first person experiences any symptoms*.[20] In fact, some research has indicated it might be as high as 79% of documented cases that stem from undocumented infectious persons in the asymptomatic or mild symptom periods.[21]

The virus is shed through particles sent out into the air (by speaking, singing, coughing, etc.) and unsuspectingly taken up into the respiratory tract of another person. There the virus readily finds the hACE2 receptor sites it needs to begin a new invasion. The odds of becoming infected are affected by a variety of factors, such as how many infected people one is exposed to, the proximity of exposure, the duration of exposure, and so on. Nevertheless, within a very short period of time—by mid-February of 2020—the highly infectious nature of SARS-CoV-2 was recognized as a global pandemic.[22]

Meeting the Enemy

COVID-19 caused by SARS-CoV-2 is an enemy to which Americans have responded in varying ways. But all of them resolve down to one of two basic alternatives: taking one's chances that either no infection will happen or if it does that it will prove of no serious consequence, *or* trying both to prevent infection and limit the likelihood of serious consequences if it does occur. To use our war metaphor, the difference is in either relying on the 'country' (i.e., one's body) to respond quickly and effectively in the event of invasion (i.e., through a natural immune response), or relying on preparing the country in advance through calling up the national guard and drafting new soldiers (i.e., preparing an immune response in advance of invasion). The latter path is taken by pro-vaccine advocates. Anti-vaxxers choose instead the former path. The thinking behind these different choices shall occupy us the remainder of this volume.

---

[19] Iacobucci, "Long Covid."
[20] Nishiura, Linton, and Akhmetzhanov, "Serial Interval." Also see Du et al., "Serial Interval."
[21] Hu et al., "Characteristics of SARS-CoV-2 and COVID-19," 148.
[22] Sanche et al., "The Novel Coronavirus."

Chapter 3
The Heart of the Debate

Clearly, when different people appeal to different sources of knowledge we can expect there may be varying conclusions. The possibility of competing conclusions is increased by differences in how matters are argued, with some people more susceptible to one kind of argument and others more susceptible to another. But we need not despair. Even if there is little possibility of ever getting everyone to agree on something (hello, flat-earthers!), there is a real possibility of achieving enough agreement on a matter to make a discernible difference in discussion and subsequent action.

As we turn now to the arguments themselves over COVID-19 vaccination we shall begin with a short effort to identify why the opposing sides have so much difficulty hearing each other. It goes back to how we *know*.

How Many Sides to the Argument Are There??

Although a bit simplistic, we might picture the situation this way:

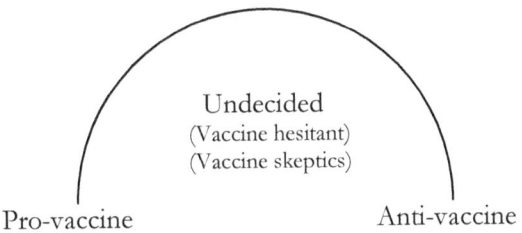

(These are not proportionally equal groups in terms of size.[23] The illustration only shows relative positions.)

Each group finds the 'heart of the matter' in a different location, and how they locate this heart is a reflection of how they think one best acquires knowledge and achieves truth. Let's consider each position on the continuum above. We will begin with the pro-vaccine position for two reasons: it has the most adherents, and the anti-vaccine position arose after it and because of it.

The Heart of the Matter for Vaccine Proponents

For proponents the heart of the matter is *a culture of practical truth*. What that means is that proponents view the situation as one where a problem exists (COVID-19) for which the desirable solution is practical in nature (vaccine), rather than ideological. That does not mean proponents have no ideology; they embrace *science* as the best way forward for solving problems in general. They envision a better, more prosperous society and richer individual lives as a result.

Those who support COVID-19 vaccination largely appeal to reason and evidence as the sources for how best to know what is true. They are willing to

---

[23] See Burki, "The Online Anti-vaccine Movement," e504.

accept limited knowledge if it proves practical to solve a problem. For them the only important debate is whether the science formed by reason and evidence is sound enough to work; they take for granted that science intends public benefit extending to all the individuals who make use of it.

This basic perspective, bolstered by a philosophy of pragmatism (truth proves itself by working!), is very American. It is part of what popularly supports capitalism, technology, and the common optimism that practical know-how can solve most any problem. Because this way of looking at life and the world is so familiar to Americans we need not spend much time here on it.

The Heart of the Matter for Anti-Vaccine Proponents

Many Americans have a harder time understanding where anti-vaxxers are coming from, so we need to spend a bit more time on their position.

Those who oppose COVID-19 vaccination reject both the popular presumption of vaccine proponents that scientists should be trusted *and* the underlying conviction that reason and evidence are the best way to know something. Instead they embrace a different, minority culture, and make popular appeals to other authorities. They propose that there are *universal, timeless truths* that should guide the individual, with society taking a backseat.

They largely appeal to authority and emotions as the best ways to know,[24] and for many this means turning to the highest authority—God, who speaks through Holy Scripture and/or some trusted religious authority. They popularly argue using forms of informal logic that presume certain shared convictions of a religious, cultural, or political nature. For them the so-called 'scientific' issues are just a thin veneer over the real issues, which are religious, cultural, and political in nature.

For so-called 'anti-vaxxers' then, the heart of the matter is an *ideological culture*, embracing both political and religious convictions. They share a vision of a culture that is at variance with that of pro-vaccine advocates, even though anti-vaxxers reserve a limited space for science. It isn't that they oppose science *per se*, but they abhor the deference given to it because to them science, when backed by government and big business, can be a bully encroaching on self-determination and liberty. For what anti-vaxxers prize most is *individual freedom*.

It may help to look at both the spheres of politics and of religion to better glimpse anti-vaccine thinking. An example related to each area may be helpful.

Politics

If we begin with politics we need first to remember that the *anti-* in anti-vaccine means a basic position formed in opposition to something else. Anti-vaxxers began as political reactionaries to government initiatives and largely continue to be better known as those who *oppose* certain things. But it is an opposition rooted in being *for* a particular vision of what liberty means.

---

[24] Davies, Chapman, and Leask, "Antivaccination Activists," 23.

With respect to political action a prominent example is the Association of American Physicians and Surgeons (AAPS), founded in 1943 to oppose universal health care. Of America's million plus physicians, less than 5,000 belong to AAPS (contrast that, for example with the American Medical Association, which has some 235,000 members). It is politically conservative, declaring on its webpage that *"Since its founding in 1943, AAPS has been the only national organization consistently supporting the principles of the free market in medical practice,"*[25] and in 2018 donated only to Republicans running for federal office.[26] This affiliation with the Grand Old Party (GOP) makes sense when we remember that Republicans, too, are better known for what they oppose than for what they stand for (e.g., limited government).

The AAPS engages in politics to support its cultural stance against a medical establishment and governmental policies it perceives as limiting the autonomy and authority of individual physicians. Thus its anti-vaccine stance is less about vaccination itself and more about it being *mandated*. Like many activist groups they use strong rhetorical phrasing and often espouse views rejected both by mainstream medicine and even by mainstream conservatism. For example, in the same way they decry Medicare as 'evil,' they liken the spectre of mandatory vaccination to 'human experimentation.'[27] In fact, they have declared immunization programs to be agents of social control reflecting a 'collectivist morality.'[28] The individual *always* comes first.

Predictably, the AAPS opposed the *Joint Statement* signed by 58 professional medical groups supporting mandatory vaccination of health care workers.[29] In their own statement, AAPS wrote:

> The Association of American Physicians and Surgeons (AAPS) declares that all human beings have the right to liberty, which they do not forfeit when they serve the sick or the disabled. The ethical commitment to protect others does not require workers to surrender their bodily integrity and self-determination and accept "the" intervention dictated by a governmental or quasi-governmental authority.

This initial declaration signals a *political* and specifically *ideological* stance (especially favoring the philosophy of Ayn Rand), rather than a medical one. Although absent of any evidence substantiating the claim, AAPS appeals to reason, offering its belief that "Long-term effects of these novel, genetically

---

[25] AAPS (https://aapsonline.org/).
[26] Khazan, "The Opposite of Socialized Medicine."
[27] See Khazan, "The Opposite of Socialized Medicine." The 2008 AAPS article "Oratory—Or Hypnotic Suggestion?" speculated that Obama was using Neurolinguistic programming (NLP) techniques to persuade voters and found it ironic that this was especially influencing Jews. The article can be accessed online at https://aapsonline.org/oratory-or-hypnotic-induction/.
[28] Seidel, "Strange Bedfellows," with special reference to an editorial published in the Spring, 2000 issue of *Medical Sentinel*.
[29] See American Public Health Association (APHA), "Joint Statement." For a full list, see the Appendix at the end of this book.

engineered products cannot possibly be known at this point. These could include autoimmune disorders, antibody-enhanced disease, infertility, cancer, or birth defects."[30]

It is important to understand what I am and am *not* saying. I am *not* saying that AAPS members publishing articles in the *Journal of American Physicians and Surgeons* (previously titled *Medical Sentinel*) are incapable of doing science. What I *am* saying is that it is neither the chief aim of the organization or of the journal to do so. The chief aim is promoting a particular cultural stance seen as more important than science. For them, science must serve, not lead the way.

Kathleen Seidel, an independent researcher known for her investigative work on claims of a vaccine-autism causal link, did an extensive study of materials published by AAPS and its journal. Her analysis found that one of AAPS' core convictions is not to embrace the concept of medical care that is itself evidence-based because it is viewed as undermining physician autonomy.[31] The right of individual doctors to act as they see fit, regardless of standard medical practice or consensus, is paramount.

Religion

Now let us turn to religion, another fundamental part of culture. The relation of science to religion has been a controversial one for a long time, despite the fact that religion was one of the sources from which science originated. Most religious people regard science as able to coexist with their religious views.[32] At best they find it confirms their beliefs, and at worse as being indifferent to them, or as operating in a separate sphere. But a minority of religious people experience science as unsettling, or even threatening, and think that a stance against it, or its commonly presumed authority, is necessary.

In the United States the religious group most negative toward science is that conservative Protestant Christian group made up of White Evangelicals. They therefore offer a second example for us to consider. Although even in this group some 57% express general compatibility with their personal beliefs and science, some 40% say their personal religious beliefs sometimes conflict (and then mostly with views about creation and evolution).[33] Again, with respect to a matter like childhood vaccination, the majority of White Evangelicals—like other religious Americans and the public as a whole—support vaccinations being required (59%). But more than a third (39%) believe this should be a matter left in complete parental control.[34] With respect to COVID-19 vaccination, even here White Evangelicals are *mostly* pro-vaccine (54%, with

---

[30] See AAPS, "Statement."
[31] Seidel, "Strange Bedfellows."
[32] In 2015, less than a third (30%) reported their personal religious views were in conflict with science, though they presume this may be the case for *others* (59%). See Funk and Alper, "Religion and Science," 4.
[33] Funk and Alper, "Religion and Science," 5–6.
[34] Funk and Alper, "Religion and Science," 28.

17% already having been at least partially vaccinated by March, 2021). But they are the religious folk most likely to oppose the vaccine (45%), too.[35]

Just as most physicians are pro-vaccine, but some oppose it, so most Christians, including White Evangelicals, are pro-vaccine, but some oppose it. We saw why anti-vax doctors might be opposed, but what about White Evangelicals?

Interviews with evangelical ministers conducted by the *Washington Post* reported their agreement that "views on coronavirus vaccines are largely shaped by political and other cultural beliefs, with government mistrust being a key factor."[36] In other words, specifically religious beliefs seem *not* to be the reason for opposition. Instead, religious ideals are attached to political or other cultural convictions.

An example of this kind of anti-vax Evangelical response is seen in the experience of J. D. Greear, President of the Southern Baptist Convention (the largest Evangelical denomination), when he announced he had been vaccinated. The announcement quickly drew more than 1100 Facebook comments, ranged across the spectrum from approval to disapproval. Among the latter were comments calling the vaccine satanic and suggesting Greear was complicit in government propaganda.[37] In short, there was a mix of religious ideas with other cultural considerations.

While we must refrain from generalizing from one example, it seems a common perception that fueling Evangelical anti-vax thinking often is *not* religious doctrine or biblical belief—or even specifically health issues. Instead, as religious studies scholar Heidi Campbell observes, it stems from conservative political affiliation with the GOP, such as identifying with ideas of smaller government, less intrusion of government in personal choices, and personal freedom.[38] In this respect, politics and religion for some White Evangelicals have become so closely entwined that separating them seems impossible. Religious conviction can be a mask for what is really political ideology.

The Culture of a Communal Sub-Culture

For both those driven by religious or political ideals (or fears) there is a shared underlying force of substantial power: *shared identity within a community*. Sociologist Brook Harrington reminds us that people don't seek everyone's approval, but they do seek the approval of their own reference groups, which establish group norms, such as affirming GOP affiliation by aligning with anti-vaccine resistance in the name of opposing Big Government. Thus, as seen, anti-vax sentiment quite often may have little to do with science concerns and much to do with protecting one's status within a certain community.[39]

---

[35] Pew Research Center, "Intent."
[36] Bellware and Cornejo, "How Pastors and Health Experts Are Struggling."
[37] Crary, "Vaccine Skepticism."
[38] Mercer, "Why Evangelicals."
[39] Harrington, "The Anti-Vaccine Con Job."

As we shall continue to see at various places in this book, whether the question of COVID-19 vaccination *should* be a matter decided by culture rather than science is an issue debated even among White Evangelicals. But we have seen enough for now to have a better idea of the thinking of anti-vaxxers about what is most important.

The Heart of the Matter for Hesitators and Skeptics

Finally, there are also those who refuse to either endorse or reject vaccination by adopting a skeptical stance. They don't regard any way of knowing as providing enough certainty to make them fully confident in one direction or the other. Most of these folk take a 'wait-and-see' attitude that at least professes openness to further information which might tip the scales for them (even if the bar for what might prove decisive is set unreasonably high).

For those who are neither proponents nor opponents the heart of the matter may be more ambiguous and variable. For some not getting a vaccine is simply because it is inconvenient. For others a pronounced fear of getting a shot—*any* shot—keeps them away (and since few see this as a justifiable reason, they form other reasons). Still others are complacent, figuring their personal risk is low, the odds of serious symptoms remote, and so no action is really needed. These all might be lumped under the general category of *personal factors* and they merit consideration even if they often aren't much part of the fierce debate over COVID-19 vaccination. In fact, as already noted, on at least some occasions personal reasons may be hidden behind one or more anti-vax objections simply because the individual imagines these will be more acceptable to a listener.

James Aten, a Christian psychologist at the prominent evangelical institution Wheaton College, has speculated that some religious folk who hesitate are motivated by what he terms a "spiritual bypass"—they defer a decision on a matter like vaccination while seeking a spiritual answer. He argues that while it is not a faith issue, faith is often appealed to in trying to respond to it.[40] But when a clear-cut answer is not at hand, then vacillating may occur.

Vaccine Hesitancy vs. Vaccine Denialism

Vaccine hesitancy may seem like a rather innocuous matter. However, in 2019, the World Health Organization (WHO) identified vaccine hesitancy as one of ten threats to global health. WHO stated, in part:

> Vaccine hesitancy – the reluctance or refusal to vaccinate despite the availability of vaccines – threatens to reverse progress made in tackling vaccine-preventable diseases. Vaccination is one of the most cost-effective ways of avoiding disease – it currently prevents 2-3 million deaths a year, and a further 1.5 million could be avoided if global coverage of vaccinations improved.[41]

This position was *before* COVID-19 became the worldwide scourge it now is.

---

[40] Bellware and Cornejo, "How Pastors and Health Experts Are Struggling."
[41] World Health Organization, "Ten Threats."

A so-called "5C" model developed from studies conducted in high-income nations has been used to characterize what drives vaccine hesitancy. The factors seem to be: confidence, complacency, convenience (or constraints), calculation of risk, and collective responsibility.[42] Any one of these five might explain an individual's hesitance to get vaccinated, and sometimes more than one factor may be in play. In addition, what is a factor at one point may give way to another factor at a different time.

Interestingly, low-income and middle-income nations apparently are more accepting of vaccination than higher-income nations. The averaged rate of acceptance across studies for such nations was 80.3%. By contrast, in the United States it was just 64.4%. The difference in rates might be explained, at least in part, by Americans with their greater access to information being inundated with information that is complicated, often incomplete, and too frequently misleading—all in service of conflicting claims.[43] No wonder, then, so many hesitate. Who is to be believed?

As problematic as vaccine hesitancy is, vaccine denialism may be more worrisome, if for no other reason than because of its influence on the hesitant. *Denialism* is the use of rhetorical speech designed to preserve the veneer of legitimate debate when, in fact, there is none. The intention is not winning an argument—because there is such a strong consensus of expert opinion opposed (which has won the assent of the public) that 'winning' has already been decided. Instead, the goal is undermining confidence so as to convince as many as possible to reject the winning argument. Why? Researcher Lucas Stolle and colleagues conclude that the motivation may be personal eccentricity or idiosyncrasy, or ideology, but it also may be greed (profiteering from a group's sentiment on the vaccine).[44]

Trust

Separating out why advocates for a position adopt the stance they hold and argue as they do is not often a simple matter. But *trust* seems to be a major factor in the level of confidence people have in the source or sources of knowledge they rely upon and the decisions they base upon the same. In short, the issue of trust is an appropriate one to round out this consideration of how people think they know something, how they then choose to argue, and what they believe is the heart of the matter being considered.

Trust vs. Mistrust of Medical Science

Those who support vaccination have more confidence in medical science than those who do not. In fact, mistrust of medicine is an often-cited reason for

---

[42] Machingaidze and Wiysonge, "Understanding COVID-19 Vaccine Hesitancy,"1338.
[43] Machingaidze and Wiysonge, "Understanding COVID-19 Vaccine Hesitancy,"1338–39.
[44] See Stolle et al. "Fact vs Fallacy," 4484. They contrast the denialist with the 'fair-minded skeptic.'

resistance to vaccination.[45] By 'mistrust' I refer to a general sense of unease, rather than the 'distrust' that might arise from some specific experience.

Mistrust, whether pronounced or trivial, does not flow from just one source. For example, it has been well-documented that mistrust of the medical care system is more pronounced among African-Americans than White Americans and reflects a history of discrimination and sometimes dramatically unethical treatment (e.g., the Tuskegee Study of the 1930s–1940s). Thus a Black American need not have personally experienced something negative to create *dis*trust, but may just have a general *mis*trust.

At least some mistrust for many Americans has come from decades of medical practice when physicians were taught in medical school to project an authoritarian confidence to patients. But this too often led to doctors not listening well to their patients and showing a profound reluctance to admit either the limits of medical knowledge or their own errors. As a consequence some people formed a defensive mistrust leading to reluctance to see a physician and/or skepticism over diagnosis and treatment, even apart from any specific experience.

Some have suggested that mistrust can stem from a sense that religion and medical science inevitably conflict at certain points, and that religion should be trusted instead of science. This seems especially true among some (but by no means all) conservative Christians, most especially groups like Christian Scientists.

With respect to vaccines, specifically parental resistance to child vaccination, both hesitancy and resistance have been found to correlate with several factors: younger age, less education, greater religiosity, greater disgust sensitivity (i.e., the frequency and intensity of experiences of disgust), and lack of trust in physicians and medicine. In terms of this last factor, it seems a chain of events often occurs: parents mistrust their pediatrician, leading them to search the internet, thereby encountering anti-vaccine information, and thus becoming vaccine hesitant or anti-vax themselves. It is easy enough to see how this might happen, when it has been calculated that 43% of sites found on top search engines for terms like 'vaccination' and 'immunization' are anti-vax in orientation.[46] (Many people are unaware that simple frequency of particular results or the order in which they are presented has nothing to do with accuracy but instead with skill at search engine manipulation.)

We saw earlier that among religious Americans, White Evangelicals are the most likely to adopt anti-vaccine sentiments, and they may express mistrust of medical science and doctors as part of a more general mistrust of science. For example, Evangelical pastor Nathaniel Manderson, who confesses difficulty in understanding how so many of his fellow Evangelicals resist exposing themselves to materials that might challenge them, sees this situation as a matter

---

[45] Stolle et al. "Fact vs Fallacy," 4483.
[46] Reuben, Aitken, Freedman, and Einstein, "Mistrust," 2. The 43% figure comes from Davies, Chapman, and Leask, "Antivaccination Activists."

of *fear*. "It's really about fear among the Christian faithful when they turn away from science," he says. He traces this fear back to a history of science reducing the need of God to explain things, producing a worry that science will somehow prove God does not exist. He interprets Evangelical resistance to COVID-19 vaccination as a response to the wider question of whether one should look to God or to science for truth.[47]

Similarly, psychiatrist Ronald Pies suggests fear or anxiety is a critical factor. He tells us we should remember that reason does not act alone in persuading us to a position; *passion* plays an important role, too. He points to research by neurologist Robert Sapolsky that finds human beings are predisposed to hear what they want to hear, and once they decide what that is they can't be reasoned out of it into something else that they were never reasoned into in the first place! But Piels also tells us Sapolsky notes the stiff price accompanying a lack of curiosity and openness—increased anxiety.[48] It is anxiety and fear that often underlie the conclusions people reach and then cling to so ferociously.

Variability in Social Trust

Mistrust of medical science and of physicians may also reflect the role of *social trust*—our confidence in the trustworthiness of others. Obviously this extends beyond such trust or mistrust in physicians (or medical researchers).

Pew Research Center polling data during the COVID-19 pandemic reveals how social trust is linked to complex interactions among personal emotions in response to the pandemic, perceptions of how different groups respond, and demographic factors (e.g., age, education, and income level). Thus, for example, older Americans with more education and higher income are more likely to have experienced less negative emotional reactions and to judge more positively the performance of others. Poll data differentiates people into three groups: high, medium, and low trusters.

Compared to polling in 2018, since the pandemic began the percentage of those showing high social trust has risen (from 22% to 29%), though the percentage of low social trusters has remained the same (35%). The belief that most of the time people will try to help others (altruism) also rose among Americans—from 37% to 42%.

Perhaps counterintuitively, high trusters compared to low trusters both believe the COVID-19 pandemic to be a significant crisis (72% compared to 63%), and to report their own lives have been changed in a major way (48% compared to 41%). Yet high trusters are less likely to report recent experiences of anxiety (39% versus 39%), sleep difficulties (22% versus 41%), or depression (16% versus 33%).[49]

In other words, as we might expect, it seems those low in social trust are like those who mistrust physicians in being more likely to experience anxiety or

---

[47] Manderson, "Evangelicals."
[48] Pies, "Anti-vaxxers."
[49] Rainee and Perrin, "The State of Americans' Trust."

other negative psychological states. We all do better when we have more confidence in who we can trust—and then actually do so. It may be that some individuals adopt passionate convictions as a way of defending against experiencing anxiety; if we can convince ourselves we are right, anxiety can be barred from consciousness (even if it lurks beneath). This also would seem to help explain a conviction of the importance of individual liberty because a mistrust of others can invite turning oneself and a few like-minded others into the only ones thought worthy of trust.

Trust Bias

A final consideration about trust is important here and stems from our previous thought. *Trust bias* is our tendency to trust most—sometimes exclusively—those most like ourselves, regardless of how trustworthy those others may be in one or more respects. Thus, once we identify with a group, any challenge to that group's positions may feel like a personal assault and prompt us to draw closer to other members of the group in a 'circle the wagons' defensive posture.

Trust bias makes it very difficult for opposing sides to hear one another, let alone change minds. This kind of bias is known in psychology as *motivated bias*, that is, "biases reflecting 'myside bias' or 'wishful thinking.'"[50] As reporter Alia Dastagir puts it, "Trying to convince someone that a deeply held view is flawed is an uphill battle."[51]

This sobering reality is underscored by research studies. For example, in one study participants were exposed to two articles, one purporting to show the effectiveness of capital punishment and the other its ineffectiveness. Not surprisingly, participants found convincing the article that agreed with their preexisting opinion, and more readily found flaws in the other. In fact, those preexisting beliefs were strengthened and became more polarized by the conflicting evidence.[52]

Should we simply resign ourselves to never being able to persuade the persuaded? No. There is ample evidence in everyday life that people do change their minds. What seem most important are two things: how they justify to themselves doing so, and how likely that justification will be accepted by others whose opinion matters to the person changing her- or his mind.

Our willingness and ability to encourage open-mindedness, and show it ourselves, may encourage others to speak honestly, listen respectfully, and perhaps change minds. We need to resist superficial notions that clinging to a misguided conviction is somehow superior to moving to a better understanding and a new conviction. The latter isn't being wishy-washy; it is being *wise*.

---

[50] Hahn and Harris, "What Does It Mean to Be Biased," 45.
[51] Dastagir, "Facts Alone." The same appears to hold true with vaccine conspiracy theories—once established in a person's mind they seem difficult to remove (Jolly and Douglas, "Prevention Is Better than Cure.").
[52] Hahn and Harris, "What Does It Mean to Be Biased," 55.

Chapter 4
The Pro-Vaccine Arguments

Proponents in favor of being vaccinated appeal primarily to science. That means they rely on a partnership of reason and evidence. Proponents popularly confer a higher authority to figures who base their claims on science.

Pro-vaccine arguments offer the following key contentions:

1. COVID-19 is a serious health emergency requiring immediate and widespread attention.
2. COVID-19 vaccines result from well-established practical science.
3. COVID-19 vaccines have been developed appropriately.
4. COVID-19 vaccines work.
5. COVID-19 vaccines are safe.
6. Unvaccinated people are at greater risk than vaccinated people are.

1. COVID-19 is a serious health emergency.

The virus that causes the disease COVID-19 is a coronavirus identified as SARS-CoV-2. Genetically less than 30,000 letters in code, its genes have been identified for 29 functions, among which are self-replication and immune response suppression.[53] The biology of the virus and the effects it produces in the disease were covered in an earlier chapter. Here we will focus on the issue of whether or not it constitutes a serious health emergency.

As unlikely as it might seem now, when the COVID-19 pandemic was in its early stages many people were influenced by anti-vaxxer claims that the virus was no more dangerous than the seasonal flu (see next chapter). The mortality rate from COVID-19 put a substantial bite into that argument.

By September, 2021 few people were indulging the fantasy that COVID-19 is 'just like the flu.' Despite massive efforts to provide vaccination—typically at no cost to recipients—as autumn dawned the pandemic levels were again rising to levels comparable to the peaks in the preceding winter season. What continued to fuel the pandemic was infection among the unvaccinated. They were disproportionately represented in infections, hospitalizations, and deaths, leaving health care officials frustrated and angry at preventable illnesses.[54]

Especially worrisome as school resumed in 2021 was the dramatic rise in pediatric cases tied to the delta variant of COVID-19. Consistent with what was being found in the adult population, a strong positive correlation exists between being unvaccinated and serious cases of the disease. Simply put, the lower the vaccination rates in an area, the higher the pediatric hospitalizations. Conversely, the higher the vaccination rates, the lower the hospital admissions.[55]

---

[53] Lehrer and Rheinstein, "Human Gene Sequences in SARS-CoV-2," 1633.
[54] Hollingsworth, Bussewitz, and Long, "COVID-19 Cases Climbing."
[55] Ducharme, "The Delta Variant." See the Figure on p. 10.

According to the COVID-19 Dashboard maintained by Johns Hopkins University, as of September 3, 2021, the pandemic had infected 219,473,393 people, leading to 4,547,932 deaths worldwide. In the United States the numbers were 39,640,020 infections with 644,468 deaths (and just in the last 28 days, 4,108,796 cases with 28,224 deaths).[56] By the middle of September the number of deaths attributed to COVID-19 in the United States was more than 675,000—a number exceeding the estimated American dead from the infamous 1918 influenza pandemic (one of history's great killers with more than 50 million dead worldwide).[57] These are staggering numbers not easy to minimize, though the last bit of information shows that COVID-19 is comparable to influenza in respect of its ability to produce a deadly pandemic. However, compared to influenza, which claims up to 650,000 deaths a year worldwide, COVID-19 is estimated to be *possibly 10 times more deadly*.[58]

Virtually every day brings reminders of how serious the situation remains. The delta variant, far more infectious than the alpha variant, which in turn was more infectious than the original strain, raises present concerns. The delta variant is so infectious that one easily finds confirmed reports such as that of an elementary school teacher who in a three-day span was the origin source of *26* confirmed infections. Of the teacher's 24 students, 12 tested positive for COVID-19—an overall rate of 50%—but for those students in closest proximity (the first two rows in the classroom), the infection rate was 80%.[59] Single anecdotal cases do not 'prove' anything, but they can be illustrative—and they certainly draw attention.

Now there are justified fears of what further mutations might bring. For example, the R.1 variant in the early autumn of 2021 was raising new concerns because it might be even more infectious than the delta variant, and already has been detected in 35 countries and two U.S. territories, as well as within the United States.[60] Without better and unified efforts to meet the enemy effectively through vaccination, pro-vaccine advocates warn, a bad situation can get worse.

For the most part, as time has gone on, anti-vaxxers have retreated somewhat from minimizing the seriousness of the pandemic by instead emphasizing more the alleged dangers of taking the vaccine and/or the desirability of various alternatives to vaccination in response to COVID-19.

2. COVID-19 vaccines result from well-established practical science.

The idea of vaccines is a simple one: expose the body to a relatively harmless or completely benign substance (the vaccine) that provokes the

---

[56] The COVID-19 Dashboard can be accessed online at https://www.arcgis.com/apps/dashboards/bda7594740fd40299423467b48e9ecf6.
[57] ABC News, "COVID Death Toll."
[58] Maragakis, "Covid-19 vs. the Flu."
[59] Lam-Hine et.al., "Outbreak."
[60] Lock, "Dangerously Mutated R.1 COVID Variant."

immune system to a response that prepares it to deal with a harmful, even deadly substance (the disease).[61]

Anti-vax arguments tend toward oversimplification and generalization by treating all COVID-19 vaccines as though they are the same. But they are not. In response to the pandemic a wide variety of delivery platforms came under development, including inactivated vaccines, recombinant protein vaccines, live-attenuated vaccines, viral vector (adenovirus) vaccines, DNA vaccines, and mRNA vaccines.[62]

How Approved American COVID-19 Vaccines Work

Neither are the currently approved vaccines for use in the United States all the same. Both the Pfizer/BioNTech vaccine (usually just referred to as 'Pfizer' and formally designated BNT162b2) and the Moderna vaccine are mRNA (messenger ribonucleic acid) vaccines. That means they rely on the science of genetics, a complicated science for most of us to grasp and therefore easily prone to misrepresentation. One anti-vaxxer claim, for instance, is that such a vaccine will rewrite the recipient's genetic code. That is a grossly misleading and completely inaccurate claim. What it actually does is provide genetic instructions to teach human cells how to produce a spike protein like that found with the COVID-19 virus. The immune system recognizes these proteins as 'foreign' and unacceptable. In response antibodies are formed. This produces an immune response like one would have to the virus itself, though with the obvious advantage of not actually having COVID-19. This means if the actual virus is encountered, the body has its defenses already prepared for action.[63]

Let's try to picture this as simply as possible for an mRNA vaccine. To do so we must enter the realm of microbiology again (see chapter 2). The mRNA vaccine 'message' helps the immune system to recognize the kind of viral profile COVID-19 presents. This is why vaccines are often spoken about as 'teaching' the immune system. An mRNA vaccine teaches the cell how to make a protein like the S protein of the virus so that the body will recognize it and generate antibodies against it. The mRNA vaccines have a very specific—and limited—affect in the body. Their limited presence means they break down harmlessly once their job is done.

A tremendous advantage is provided by vaccination over relying on infection to achieve 'immunization.' In both cases a human body begins with *no* antibodies to the SARS-CoV-2 virus that causes COVID-19. The vaccinated body, though, preemptively begins making antibodies so that if and when the

---

[61] Science and Technology professor Stuart Blume (*Immunization*, 14) puts it this way: "Vaccines usually contain antigens, which are bits of the disease-causing organism, which stimulate the body's immune system to fight off a potential infection. The vaccine helps the immune system to recognize the pathogen and to launch the right kind of counter-attack before the sickness has taken hold."

[62] Liang, et al., "Adjuvants," 1.

[63] A more developed explanation can be found on the CDC website; see CDC, "Understanding mRNA COVID-19 Vaccines."

virus is encountered, *the body is already prepared to fight*. The unvaccinated body must react, responding on-the-fly, with no previous preparation.

While the mRNA vaccines require two doses, the Johnson & Johnson (J&J, formally designated *JNJ-78436735* or Ad26.COV2.S) vaccine is a single dose vaccine—an attractive alternative to those who strongly dislike shots. The reason the J&J vaccine requires only one shot is because it is a different kind of vaccine. It is a Viral Vector vaccine, meaning it employs a virus—the so-called 'vector'—to prompt an immune response. The virus used is *not* COVID-19. But it is one that tells the cells it invades to make a spike protein like that found with COVID-19. When the infected cells show this protein the immune system reacts by producing antibodies *as if* the body was facing the more serious virus. In this manner, without getting COVID-19, the body learns to recognize it and is ready if it encounters it.[64]

Vaccine History 101

Safe vaccines have a long history. Vaccination efforts are historically tied to medical responses to smallpox. Such efforts can be traced back as far as 16th century China.[65] However, most people are more familiar with the work of Edward Jenner (1749–1823), who administered his first smallpox vaccination in 1796.[66] That was a long time ago and provides a rich historical backdrop.

We now possess a history of the science of vaccination extending back more than 200 years—and more than 300 years when we add in variolation (treating smallpox by using smallpox itself). Despite the criticisms of anti-vaxxers, vaccination science has persisted and expanded both because it has been needed and because it has succeeded. It is hard to imagine that vaccination would have been so much further developed and extended to so many diseases apart from a reliably good record of success leading to widespread acceptance.

From the start vaccines have served a concrete, practical scientific interest and one aimed at general welfare. This 'general welfare' has extended beyond people to animals; human vaccination and veterinary vaccination share close ties as well as a history tracing back to the smallpox vaccine (which used cowpox).[67] Whether or not we consider the extension to animal care, a broad point is worth emphasizing: vaccines are not just about individual health but also about public health. This is obvious in their key functions.

Three Important Functions of Vaccines

As they have developed, vaccines have come to serve three important functions: first, they reduce the probability of infection (and in a best case scenario entirely prevent infection); second, they reduce the severity of symptoms; finally, they also reduce the rate of transmission of the disease. The

---

[64] See CDC, "Understanding Viral Vector COVID-19 Vaccines."
[65] Leung, "'Variolation' and Vaccination," 5.
[66] Baxby, "Edward Jenner's Role," 13.
[67] Lombard, Pastoret, and Moulin, "A Brief History," 30. Vaccination as such began with Jenner's use of cowpox, which proved as effective as variolation but was demonstrably safer.

last function promotes *herd immunity*—an indirect benefit of enough people becoming immune or resistant to a disease so as to severely inhibit its spread.

Obviously, two of these functions work at the level of the individual, while the third also works at the community level.[68] This third function—the *social benefit* of vaccination—has become a lightning rod in the debate over vaccines. That is because as vaccinations succeed individual risk of infection is lowered and some people 'piggyback' off the risk others have taken to enjoy the benefits of widespread immunization. Thus, relatively speaking, the private risk/benefit ratio is lower than the social risk/benefit ratio. Put another way, while it is clearly in the *public* interest to have widespread vaccination (low risk/high benefit), for any given *individual* the matter may be less clear and so motivation to get vaccinated may diminish.[69]

We actually have seen that with the COVID-19 pandemic. As more people became vaccinated and infection numbers dropped, fewer people chose vaccination—until infection numbers rose again. This is reflected in the so-called 'waves' of the pandemic. The more direct the threat is, the higher the motivation to act. All of this focuses attention on the role of ethics, specifically the ethics of personal freedom. But that shall await a later chapter.

The Economics of Vaccines

Because making and distributing vaccines carries an economic cost, societies as well as individuals weigh cost/benefit ratios. Generally and typically, in response to epidemics and pandemics the general population and their governments want vaccination to occur, despite its costs, because the economic and health costs of not doing so are greater than the costs associated with production and delivery of vaccines.

We sometimes hear of the reluctance of pharmaceutical companies to make vaccines. A *patented vaccine* can make money for its creator because there is a lack of competition. Absent patents there can be enough competition that vaccine prices have to stay low and thus profit margins may be minimal. In such a setting basic capitalism prevails and pharmaceutical companies invest little interest and show little effort to research and create new vaccines or involve themselves with existing ones.

With the COVID-19 pandemic this environment was dramatically changed. The seriousness of the threat was met by unprecedented response by governments and pharmaceutical companies. The former were motivated both by a desire to safeguard citizen well-being and by a desire to protect economies. (Which of these two has been more of the priority of governments is quite debatable, but a matter for discussion elsewhere.)

Anthony McDonell, a policy analyst, and Flavio Toxvaerd, an economist—whose discussion of the economics of COVID-19 vaccination I am largely

---

[68] Blume, *Immunization*, 19, points out that vaccines aim to facilitate two kinds of health—that of the individual body and the 'social body,' the community.
[69] McDonnell and Toxvaerd, "How Does the Market for Vaccines Work?"

following—observe that many governments were willing to take steps to 'de-risk' vaccine development for pharmaceutical companies to incentivize full-hearted efforts. This does *not* mean that safety protocols were set aside. As they explain, "In other words, governments stepped in and committed to fund the research, cover costs and buy vaccines at agreed prices so that pharmaceutical companies could pursue vaccine candidates with uncertain prospects without worrying about the financial consequences."[70]

We might do well to remember that pharmaceutical companies are not just about profit. Many people who work for such companies—including those in power within them—genuinely believe in contributing to the public good. McDonell and Toxvaerd point out that even with making a profit the *value* of the money made remains *below* the social benefit derived from their vaccines. They know that for every person directly benefited from receiving the vaccine there are others indirectly benefited by that action. In this manner, as observed earlier, vaccines contribute to herd immunity. Nobel prize-winning economist Michael Kremer has estimated the benefits from COVID-19 vaccines are somewhere between *40–300 times greater than the price being charged* by the pharmaceutical companies.[71]

Medical Profession Support

The medical community is pretty sold on the benefit and safety of COVID-19 vaccines. Yes, there are so-called 'anti-vax doctors,' but they are both in a small minority and they appear principally motivated by cultural ideals (e.g., complete autonomy for physicians), or a profit motive (i.e., they make a substantial income from anti-vax activities). Most doctors have been emphatically in favor of vaccination.

In fact, the Federation of State Medical Boards (for general information, see https://www.fsmb.org/) opened its statement on doctors spreading anti-vax misinformation this way: "Physicians who generate and spread COVID-19 vaccine misinformation or disinformation are risking disciplinary action by state medical boards, including the suspension or revocation of their medical license." That is not because the Federation is a draconian government agency hell-bent on suppressing free medical practice, but because—as it explicitly states—doctors are vouched strong public trust they have an ethical obligation to protect by their practice of medicine and communication of what they do, both being based on consensus-established scientific facts. Spreading misinformation is an irresponsible ethical violation.[72]

---

[70] McDonnell and Toxvaerd, "How Does the Market for Vaccines Work?" Also see Chen and Toxvaerd, "The Economics of Vaccination."
[71] McDonnell and Toxvaerd, "How Does the Market for Vaccines Work?"
[72] Federation of State Medical Boards (FMSB), "FMSB: Spreading COVID-19 Vaccine Misinformation."

3. COVID-19 vaccines have been developed appropriately.

A very large concern, especially early in the pandemic, was that COVID-19 vaccines were being rushed in their development. This concern was still present in early 2021 when some two-thirds (67%) of those who said they definitely or probably wouldn't get the vaccine named it as a reason why.[73] It has been routinely pointed out that the previous 'record' for vaccine development was that for mumps, which took about four years. How, then, could a safe and effective vaccine be produced in less than a year?

It is a fair question, and one proponents have conscientiously answered from the beginning. They point out, first, that some historical context is needed: the mumps vaccine was developed more than a half-century ago. Think about how much scientific and technical advancement has occurred since then. Computing power that once required a large room now fits on a watch.

Second, with specific reference to the current pandemic, the antecedent research for COVID-19 vaccines is almost a century old. Coronaviruses have been known since 1930 when so-called 'infectious bronchitis virus' (IBV) was identified in chickens. By 1968 this and related viruses were being reclassified as 'coronaviruses'—a name derived from its appearance with distinctive spike proteins. In 2002, because of the Severe Acute Respiratory Syndrome (SARS) epidemic, this coronavirus' genome was decoded.[74] The appearance in 2012 of Middle Eastern Respiratory Syndrome (MERS)—another coronavirus—sparked renewed research attention. Physician Peter Hotez, professor of pediatrics and molecular virology and microbiology at Baylor College of Medicine, relates that the research group he is part of laid the groundwork for the current vaccines with their work over the last decade. Specifically, he elaborates, "We learned that the spike protein is the soft underbelly of the virus. And we showed that if you deliver the spike protein as a vaccine, it's highly effective—induces what are called virus-neutralizing antibodies."[75] Anti-vax criticisms ignore this background.

Third, mRNA vaccines like Pfizer's and Moderna's benefit from an approach that facilitates more rapid development and testing than previous vaccines required. It is, in effect, a more efficient model, more readily susceptible to current technology, and more reliable in its effects.

Fourth, a series of factors coalesced in an unprecedented way as part of the intentional rapid-response demanded by the American people. These included massive infusions of public money, international collaboration among scientists and organizations, working out a practical balance between regulatory steps and authorization, and utilizing already existing research into related coronaviruses.[76]

---

[73] Funk and Gramlich, "10 Facts about Americans and Coronavirus Vaccines."
[74] For an expanded and very accessible history of this sequence, see Fauzia, "Fact Check."
[75] Berg, "How a Decade of Coronavirus Research Paved Way."
[76] These factors are discussed in numerous places, including in the Evangelical Christian "Christians and the Vaccine" FAQs page.

The response of the pharmaceutical industry proved remarkable. According to the World Health Organization (WHO), as of August 25, 2021, there were 184 vaccines being explored in laboratories and with animals. Some 37 were being tested with healthy young individuals (Phase 1), 36 with broader test populations (Phase 2), 31 in large international trials (Phase 3), and 20 are presently in use, with 8 being monitored after approval (Phase 4).[77] Most Americans, familiar only with Pfizer, Moderna, and Johnson & Johnson, have little idea of the efforts being made globally.

4. COVID-19 vaccines work.

A major contention of pro-vaccine advocates is that *vaccines work*. They often appeal to the history of vaccines to substantiate this claim. For example, medical reporter Tish Davidson, in her history of vaccines, points to the success against a disease little noted today. "In the 1920s," she writes, "before an effective diphtheria vaccine became available, at least 200,000 people in the United States contracted the disease each year. Between 10,000 to 20,000 of them died. The death rate in children under five was 20 percent. Now sometimes years go by without a single case in the United States."[78]

The same situation prevails for many diseases. Yet this very success has had an unintended consequence. People who grow up in a world without smallpox, with minimal risk of polio, or measles, or diphtheria, or a host of other serious illnesses, tend to discount the seriousness of these diseases or the risk of their recurrence, and thus question or resist the espoused need for continuing vaccination. Kenneth Carmago, Jr., a medical doctor and university professor, points out that when the visible benefit of vaccination is removed a powerful motivation to protect oneself and one's children disappears.[79] (But there are still those of us old enough to remember a world where the fear, for example, of contracting measles and dying from it was fueled by knowing someone who did and this memory adds a certain context that provides motivation to vaccinate.)

Anti-vaxxers question the effectiveness of COVID-19 vaccines by asking, 'But what about so-called breakthrough infections (i.e., those occurring in the vaccinated)?' They like to highlight reports of such infections as evidence the vaccine doesn't work and, coupled with possible side-effects, is dangerous. Pro-vaccine proponents can respond by pointing to studies that indicate otherwise.

For example, a study published in September, 2021, reported that even in breakthrough infections vaccination confers important protection. The study, using information obtained from more than 1.2 million people reporting a first vaccination dose and more than 971,000 reporting full vaccination, found that compared with unvaccinated people who contract COVID-19 both the partially

---

[77] Gavi, "The COVID-19 vaccine race—weekly update (25 August 2021). Similarly, Machingaidze and Wiysonge, "Understanding COVID-19 Vaccine Hesitancy," 1338, note that around the world there are now more than 125 vaccine candidates, 365 vaccine trials ongoing, and 18 approved (by at least 1 country) vaccines.

[78] Davidson, *Vaccines*, xi.

[79] Camargo Jr., "Here We Go Again," 2.

and fully vaccinated had reduced risk of hospitalization, with fully vaccinated people also having a much reduced risk of long-term symptoms.[80]

More specifically, in studying 6030 people who contracted COVID-19 after one dose of vaccine and 2370 people who contracted the virus after full vaccination, they found that after a second dose the odds of showing long term symptoms (i.e., 28 days or more) were approximately halved. Compared to the unvaccinated sick, infected vaccinated people as a group showed more who were asymptomatic, more who had fewer symptoms, and a far reduced risk of hospitalization.[81] Vaccination does not promise *no risk* of infection but offers a *significantly reduced risk*. Moreover, when infection does occur, compared to infection in the unvaccinated the risk of many and serious symptoms, or death, is dramatically reduced.

5. COVID-19 vaccines are safe.

Let's pursue that last idea. Not only do vaccines work, they are safe. However, "safe" does not mean "absent of risk." When we call playground equipment safe we don't assume that means no one can get hurt on it. The same is true of cars and a great many other things we can think of. Similarly, medical treatments, including drugs, carry some risk. By "risk" is meant an adverse event plausibly linked to the treatment.

Vaccines carry such risk but remain safe because the risk is so low and, generally, the consequences minor *relative to the risk and consequences of the disease itself*. As always, context matters. Simply speaking about vaccine risk absent a comparison to the risk of contracting the disease is disingenuous and irresponsible.

The risks people express concern about, and that anti-vaxxers try to emphasize, are side-effects associated with receiving the vaccine. Allergic reactions, though rare, can be serious. The CDC and FDA examined such reactions to the Moderna vaccine in early 2021. Of 4,041,396 first doses there were reported 1,266 adverse reactions (0.03%), 108 of which indicated possible *anaphylaxis* (a possibly life-threatening allergic response); of these 108 possibilities, 10 were found to be anaphylaxis.[82] While 10 such events are concerning—especially to those who experience them (and as someone with a lifetime of serious allergies I don't minimize this!)—they represent a miniscule risk (less than 0.00025%).

An individual can, of course, bet that they won't ever contract COVID-19 and thus by avoiding vaccination they can eliminate any risk associated with it. Some people do exactly that—bet their health, possibly even their life, on being able over a prolonged time to avoid infection. All pro-vaccine adherents can say to that choice is, "Best of luck."

---

[80] Antoneli et al., "Risk Factors."
[81] Antoneli et al., "Risk Factors," 8.
[82] CDC/FDA, "Allergic Reactions."

6. Unvaccinated people are at greater risk

Pro-vaccine proponents can argue 'the proof is in the pudding'—that is, the actual, demonstrable results of COVID-19 vaccination substantiate the claim that the vaccines work, that they are safe, and that they substantially lower risk.

Vaccine proponents argue careful documentation shows that the risk of hospitalization and death decline dramatically after full vaccination. Thus the CDC reported that as of August 23, 2021, *more than 171 million* Americans had been fully vaccinated and yet there had been reported only 11,050 individuals with so-called 'breakthrough' infections resulting in non-fatal hospitalization (8,987) or death (2,063; i.e., 23% of those hospitalized).[83] In other words, hospitalization was very rare and death even rarer—a mere 0.0012% (based on 171 million people). Or to put it differently, a fully vaccinated American has a far greater chance of being struck by lightning (1-in-15,300) than dying of COVID-19 (1-in-100,000), *even if infected.*

Anti-vaxxers like to point to raw numbers—and 2,063 fully vaccinated dead Americans is a scary number. But context is everything and knowing that while vaccination provides no guarantee it is reassuring that it increases the odds so dramatically in one's favor—especially compared to the unvaccinated.

Of course, as the CDC is up front in acknowledging, the number of *reported* cases is almost certainly fewer than the number of *actual* cases. But simple math says the actual number would have to be dramatically larger to much reduce the favorable odds—and there is no evidence that such is the case. (Interestingly enough, anti-vaxxers sometimes argue that instead of being undercounted, COVID-19 numbers are exaggerated. But that idea will have to wait until the next chapter.)

A principal argument in favor of vaccination follows the observation from Nature that viruses mutate when allowed to freely reproduce across numerous hosts. Thus the delta variant of COVID-19 was enabled to arise and it has proved more transmissible than its predecessor variants. The history of the virus to this point has been one of ever increasing transmissibility without an accompanying decrease in serious health effects.

The seriousness of the situation is reported in scientific studies. For instance, a study of more than 43,000 confirmed cases of COVID-19 in England found that "patients with the delta variant had more than two times the risk of hospital admission compared with patients with the alpha variant." The only good news in this is that the same study found "vaccination leads to a similar relative reduction in the risk of hospitalization for patients with the delta variant or the alpha variant."[84] In other words, despite the increased danger posed like variants such as delta, vaccination still confers a sizable advantage to survive and stay well.

---

[83] CDC, "COVID-19 Vaccine Breakthrough."
[84] Twohig, "Hospitalization," 5.

Chapter 5
The Anti-Vaccine Arguments

Anti-vaccine arguments are by their very name and nature *against* a position. This chapter highlights the arguments against the COVID-19 vaccines as made by anti-vaxxers. The term 'anti-vaxxers,' by the way, is sometimes used by pro-vaccine advocates as a derogatory one—almost a swear word—but here it is intended simply as a short, convenient descriptor.

Objections to Vaccination

Below is a list of common objections, in no particular order. Some are very specific to COVID-19 vaccines; others are ones often raised against many vaccines. I will address other vaccines only incidentally.

1. COVID-19 is not that dangerous.
2. COVID-19 vaccines were developed too fast and are dangerous.
3. COVID-19 vaccines were developed by Big Pharma, which only cares about profit.
4. COVID-19 vaccines are backed by Big Government, which threatens the sanctity of individual decisions about personal medical care.
5. COVID-19 vaccines include dangerous ingredients (e.g., preservative ethylmercury or aluminum salts).
6. COVID-19 vaccines include morally objectionable ingredients (i.e., aborted fetal cells).
7. COVID-19 vaccines aren't safe; they carry an unacceptable risk of undesirable side-effects.
8. COVID-19 vaccines actually increase risk of infection.
9. Other, better courses of action exist (e.g., hydroxychloroquine, ivermectin, vitamin C or D, zinc) than getting a COVID-19 vaccine.
10. COVID-19 will decline without the need for vaccines; natural 'herd immunity' is superior to vaccine immunity.

Some of the above objections—or at least concerns—are more prominent than others in what people actually say. In early 2021, Pew Research Center survey data found among anti-vaxxers the following cited reasons for saying that they definitely or probably won't get a vaccine were the most common:

- ❖ concern over side-effects (72%);
- ❖ concern over the rapid development of vaccines (67%); and,
- ❖ concern over how well the vaccines work (61%).[85]

Let us consider each objection in turn, listening to what anti-vaccine advocates themselves say. Just as we heard pro-vaccine arguments as mounted

---

[85] Funk and Gramlich, "10 Facts about Americans and Coronavirus Vaccines." Interestingly, a little more than one-third (36%) of these people also opposed vaccines in general.

by pro-vaccine advocates, so we need to be fair and to hear anti-vaccine arguments as presented by anti-vaxxers and not their opponents.

1. *COVID-19 is not that dangerous.*

One anti-vaccine voice who has garnered much attention is pharmacologist Mike Yeadon, a former Vice President at Pfizer, who proclaimed the COVID-19 virus to be less of a threat to those under age 70 than the seasonal flu. He calculated that in 2020 the disease would take about four months (February to June) to spread across the United Kingdom (UK), and judged that by March of 2020 it was already in retreat. In his view, the prediction of some 40,000 deaths in the UK was a sound one.[86]

Yeadon's response to the subsequent rise in numbers beyond what he expected, and to a so-called 'second wave,' was to flatly proclaim a second wave could not happen. He blamed reports of such a wave on the widespread use of the polymerase chain reaction (PCR) test for COVID-19, which he judged as being used in such a manner as to make it "completely worthless." In his view any claimed 'cases' are an artifact of "a deranged system."[87] In other words, COVID-19 numbers are being grossly exaggerated.

Similarly, in Germany biologist Karina Reiss and retired medical microbiologist Sucharit Bhakdi published a book titled *Corona False Alarm? Facts and Figures*, in which they argue that estimates comparing COVID-19 with seasonal influenza were invalid as reported by WHO and the CDC. They take particular issue with the CDC estimate that COVID-19 is ten times more dangerous than the seasonal flu. Reiss and Bhakdi claim that such an estimate depends on ignoring the most mild and asymptomatic cases. They offer their own estimate: a 0.4% fatality rate—one comparable to the seasonal flu.[88]

In the United States, physician Joseph Lapado argues that the real problem is what he terms 'Covid Mania'—a social condition in which one illness (and not a particularly dangerous one at that) is placed above all other problems in society. While acknowledging the pain and loss among older adults and their loved ones, he cites a low mortality rate (0.01%) for those under age 40, and concludes such numbers show COVID-19 has never posed a serious threat to either social or economic institutions. He argues that 'Covid mania' has resulted in scientific research being too narrowly framed and interpreted. As a result, he says, there has been a fixation on impractical measures to slow the spread of the disease and a call for vaccination when the vaccine risk/benefit ratio remains unknown. He thinks it foolish to believe the COVID-19 variants can ever be outrun by vaccines. For him, the bottom line is "rational decision making," by which he means cost/benefit analyses that include many other things in life that people value and not merely COVID-19.[89]

---

[86] Yeadon, "The PCR False Positive Pseudo-Epidemic."
[87] Yeadon, "The PCR False Positive Pseudo-Epidemic."
[88] Reiss and Bhakdi, *Corona False Alarm?* See chapter 2.
[89] Lapado, "An American Epidemic of 'Covid Mania.'"

A more extreme approach is represented by psychiatrist Andrew Kaufman, who denies that the COVID-19 virus (SARS-CoV-2) exists. He claims that scientists have mistaken exosomes for viruses (which don't actually exist). An *exosome* is a small sac (a "vesicle") secreted by cells; these sacs contain bits of DNA, RNA, and proteins. Kaufman argues that it is these that are taken up into other cells and cause trouble. The so-called COVID-19 virus is just a particular genetic sequence randomly generated in the human body.[90] Far more dangerous, then, are the vaccines designed to 'treat' a non-existent virus.

2. COVID-19 vaccines were developed too fast and are dangerous.

We encountered the objection that COVID-19 vaccines were developed too quickly in the previous chapter, but since it is such a common concern we can profit ourselves considering it again in slightly different fashion. This anti-vax argument advances two claims: first that the vaccines for COVID-19 were developed too quickly, and second that they are dangerous. These two claims are sometimes joined in a modified and more modest claim that at best the vaccines' effects are not yet adequately known because sufficient time for safety testing was not provided. Here are ways each claim is advanced.

2.1 The Vaccines Were Developed too Quickly

First, anti-vaxxers commonly point out that development of vaccines historically occurs at a plodding pace—sometimes decades—and that the previous record was still *four years*. Therefore, development of a vaccine in a year or less looks suspiciously rushed. They urge we not look past this fact.

With respect to coronaviruses (of which the SARS-CoV-2 virus that causes COVID-19 is one), anti-vaxxers point out that previous efforts at developing a vaccine have been plagued by problems. In an article for *Vaxxter*, journalist Rachael Parsons quotes renowned immunologist (and vaccine developer) Ian Frazer, who in an interview remarked, "One of the problems with corona vaccines in the past has been that when the immune response does cross over to where the virus-infected cells are it actually increases the pathology rather than reducing it."[91] Therefore, swiftly approved vaccines before adequate time for safety trials, especially in light of a checkered past history of such efforts, call for rejection.

But the anti-vaccine argument actually extends well-beyond appeals to history. They argue that the speed of development reflects Big Government working with Big Pharma at the expense of the Little People (i.e., average Americans). This has been a widely shared concern by anti-vaxxers and others. A survey conducted in early October, 2020, by the Kaiser Family Foundation found 62% of Americans concerned that the federal government's FDA (Food and Drug Administration) was under political pressure from the Trump administration to approve a vaccine before it was established as safe. In fact,

---

[90] See Kaufman's videos for details; cf. Jarry, "The Psychiatrist."
[91] Parsons, "Faith, Trust or Science—The COVID Vaccine, Part 1."

more than half (55%) of Americans suspected the President's direct involvement in pressuring the FDA's review.[92]

In light of these matters anti-vaxxers find claims of vaccine safety implausible and urge at the very least 'vaccine hesitancy'—a wait-and-see attitude until adequate safety trials are finished. Hard core anti-vaxxers, who reject all vaccines, simply use this matter of the quickness of vaccine approval one additional reason to reject it.

2.2 The Vaccines Are Dangerous

The second claim is that COVID-19 vaccines are dangerous. This claim flows from the previous one because having been developed too quickly the vaccines' dangerous aspects have not received adequate attention.

It is important to start with a recognition that this specific claim, like many others, is thought by anti-vaxxers to apply to *all* vaccines. In that sense, it can stand on its own and need not depend on the idea of a too rapid development of COVID-19 specific vaccines. This more general argument likes to appeal to *demonstrable instances of actual injury* linked to receiving a vaccine. Social media has provided a wide platform to share stories in this record. But anti-vaxxers also like to utilize the national Vaccine Adverse Event Reporting System (VAERS) which provides statistics and case information.

Beyond individual anecdotes of harm or grand statistics, though, anti-vaxxers want to make a broader point: *even the U.S. Supreme Court acknowledges enough dangers in vaccines as to call them unsafe*. Citing Justice Scalia's opinion in writing for the majority decision in the 2011 case Bruesewitz v. Wyeth, anti-vaxxers argue that in calling vaccines "unavoidably dangerous" they were also saying vaccines are 'unsafe.'[93]

Specific to COVID-19 vaccines, retired microbiologist Sucharit Bakhdi claims in a YouTube video both that the pandemic is "fake" and that the vaccines are deadly. He remarks, "They are forcing vaccination on people, and I believe they are killing people with this vaccination."[94]

Further, anti-vaxxers point to various studies comparing unvaccinated Americans with those who have been vaccinated. The vast majority of Americans have received a vaccination of some kind, so the completely unvaccinated population is very small. But anti-vaxxers point to scientific studies suggesting unvaccinated children experience fewer developmental delays, ear infections, and gastrointestinal disorders, and less asthma.[95] Another study found vaccinated children at far more risk for developing attention deficit hyperactivity disorder (ADHD).[96] Thus they contend that remaining unvaccinated leads to healthier children and, eventually, healthier adults.

---

[92] Lewis, "What Is Driving the Decline?"
[93] *Vaxxter*, "New Survey."
[94] Bhakdi is quoted in Funke, "Fact Check: COVID-19 Vaccines Don't Cause Death."
[95] Hooker and Miller, "Analysis."
[96] Lyons-Weller and Thomas, "Relative Incidence."

3. COVID-19 vaccines are really just about profit.

One of the most prominent anti-vaxxers is Robert F. Kennedy, Jr., who has authored anti-vax books with titles like *Thimerosal: Let the Science Speak* (2014) and *Vaccine Villains: What the American Public Should Know about the Industry* (2016). His influence, though, seems to come most notably through Children's Health Defense (CHD), of which he is founder, chairman of the board, and chief legal counsel. He makes the argument on the CHD website that the handling of vaccines undermines capitalism. He puts the current vaccine situation into what he sees as the proper context: "There's no market for vaccines because they couldn't sell them. So, they had to disable the markets and force us to take them against our will." Worse, he continues, "They have dismantled all of the institutions of our Democracy in order to make this business function."[97]

Ophthalmologist Kristen Held, President of the Association of American Physicians and Surgeons (AAPS) during the early stages of the pandemic, argues that the statistics presented by pro-vaccine groups are a means of manipulation. "The manipulation," she says, "is driven by power and money and fueled by fear mongering, panic stoking, and promise of monetary and political gain." She points to economic (and power) incentives attached to increased testing, tracing, and reporting of COVID-19. For example, she points to hospitals being paid handsomely for each COVID-19 admission.[98]

In an interview with LifeSiteNews, pharmacologist Yeadon accuses pharmaceutical companies of unnecessary production of booster shots, aided and abetted by rubber-stamping governmental agencies like the FDA—all part of an overall strategy. In his view, as he puts it, "There is no question in my mind that very significant powerbrokers around the world have either planned to take advantage of the next pandemic or created the pandemic."[99]

What all these arguments have in common is the contention that the response to *the so-called COVID-19 pandemic is driven by selfish vested-interest on the part of those who stand to profit from it*. Although Big Pharma comes under the most direct criticism, other power entities—like hospital systems—are not exempted. And because money and power are so intricately linked in American society, this argument is inevitably linked to the next.

4. COVID-19 vaccines are backed by Big Government,
which interferes with individual decisions.

In the same interview just mentioned above, Yeadon extends his accusation to governments around the world. He argues the response to COVID-19 shows not merely "convergent opportunism" (i.e., an accidental convergence of governments all being opportunistic), but an actual "conspiracy." Collectively these governments are using COVID-19 for totalitarian purposes such as controlling their populations' movements through vaccine passports. Furthermore,

---

[97] Kennedy, Jr., "Standing Up for Our Children."
[98] Held, "COVID-19 Statistics and Facts." The quote is from p. 70.
[99] Delaney, "EXCLUSIVE—Former Pfizer VP."

he maintains, Big Pharma and Big Tech are collaborating with Big Government in spreading propaganda while censoring truth. He argues that the end aim in this is a worldwide database so that people can be identified and controlled as to vaccination status. Why? "I'm very worried," Yeadon says, "that pathway will be used for mass depopulation, because I can't think of any benign explanation." Beyond that, he fears, lies the shadow of totalitarian governments in the name of eugenics cleansing their populations of those they deem undesirable, even using mass murder.[100]

This claim of *intrusive government*, more than any other one strikes to the heart of the matter for anti-vaxxers. Even before the pandemic, an analysis of some 480 anti-vax websites found that nearly 79% expressed a distrust of government.[101] Many Americans express little, if any, trust in the government providing accurate information about coronaviruses or COVID-19. In fact, a late March, 2021, Ipsos/Axios poll found 45% expressing little or no trust in government in this respect.[102]

Physician Jane Orient, head of the Association of American Physicians and Surgeons (AAPS),[103] before the beginning of the pandemic, wrote a statement on behalf of the AAPS. She makes the following argument in it[104]:

1. Patients have *personal rights*, for example, to informed decisions about medical care or making decisions for their children.
2. Forcing people to assume "government-imposed risks" violates those rights.
3. All vaccines are by their nature risky.
4. The risk-benefit ratio involved must be decided not by the federal government, but by individual patients and physicians.

Orient proposes the solution historically favored by the anti-vaccine movement: *better public health measures*. She approvingly cites a law article favoring informed consent, better communication, and market-based approaches—all ways to prioritize individual freedom and protect capitalism. However, the bottom line remains the protection of individual liberties. In her estimation, an unvaccinated person with no exposure to a disease should not have her or his freedom restricted.

In an interview, Orient expresses her resistance to government pushing risks on people. She reiterates her basic stance: respect of individual rights. In that light she asks, "Where is 'my body, my choice' when it comes to this?"[105]

---

[100] Delaney, "EXCLUSIVE—Former Pfizer VP."
[101] Moran, "Anti-Vaxx Websites."
[102] Ipsos/Axios Poll (March 19-22).
[103] The AAPS was introduced in chapter 2. Orient does not like the label 'anti-vaxxer' applied to herself though she favors their cultural position.
[104] Orient, "Statement."
[105] Orient is quoted in Stolberg, "Anti-Vaccine Doctor."

That idea—*my* body, *my* choice—resonates so much with anti-vaxxers as to serve as a rallying cry.

Florida's Surgeon General, Joseph Lapado, though, has shifted much of the blame to scientists and their influence over government policy makers. In an interview after his appointment to the state's highest health position, Lapado claimed that the public climate of mistrust is attributable to those scientists who are "taking the science and basically misrepresenting it to fit their agendas."[106] Thus, while generally distrustful of government overreach, anti-vaxxers typically support conservative politics that emphasize decisions at the local-level and that support the preservation of individual liberties.

5. *COVID-19 vaccines include dangerous ingredients.*

A common argument made against vaccines is that they contain dangerous ingredients. When such claims are left vague and divorced from context they can promote considerable fear. Anti-vaxxers are often as guilty as pro-vaccine advocates at making broad claims about vaccine ingredients. Relatively rarely are efforts made on either side to specify what ingredients are present in a vaccine, but as just mentioned, this likely stems from a sense that few people will understand such information even if it is provided. After all, most Americans presented a list of ingredients for a vaccine are left shaking their heads as to what danger, if any, they might pose simply because they don't know what those ingredients are—even if named.

Anti-vax advocate Brandy Vaughan, a former pharmaceutical representative (i.e., sales person) for Merck, on her website Learn the Risk lists 15 ingredients found in vaccines she identifies as toxic. She also offers a table of vaccines and their ingredients.[107] However, COVID-19 vaccines are *not* included.

Still, it is common to find in social media claims about harmful ingredients in the COVID-19 vaccines.[108] While it is beyond the scope of this book to examine every ingredient that might be suspected of causing harm, we do need to consider both aluminum and thimerosal, two of the most commonly cited ingredients in anti-vaccine arguments and also claimed to be in the COVID-19 vaccines.

5.1 Aluminum (Alum)

Let us consider first the ingredient *alum* (aluminum oxy-hydroxide). It has been used in vaccines since the 1930s as an adjuvant (an ingredient that enhances immune system response). The exact mechanism(s) by which the desired effect is achieved remain uncertain, and gives even scientists pause. Anti-vaxxers point out the inherent neurotoxicity of aluminum and remind us that some research finds toxic side-effects related to vaccination beyond actual allergic reaction.

---

[106] Mower and Wilson, "Florida's Next Surgeon General."
[107] Vaughan, "Do You Know What's in a Vaccine?"
[108] See, for example, the examples from social media linked to in the Reuters Staff, "Fact Check: COVID-19 Vaccines Do Not Contain the Ingredients."

5.2 Thimerosal/Ethymercury

Thimerosal raises concerns because of *ethylmercury*. Most folks' eyes gravitate at once to the 'mercury' part of the name. It is common knowledge that mercury poisoning is serious. A substance like dimethylmercury is so toxic even a little in contact with skin can be fatal. Mercury can build up in the body, too, though the majority of it is absorbed, for instance through the methylmercury encountered in eating some seafood, and then discharged from the body through breath and human waste (urine and feces).

Of course, ethylmercury is not the same as mercury, but that by itself does not make it safe. A common ingredient in multidose vaccine vials is *thimerosal*, an ethylmercury-containing preservative. It can prompt minor reactions like redness or swelling at the place where the vaccine is administered by a shot.

6. COVID-19 vaccines include morally objectionable ingredients.

Some people reject supporting COVID-19 vaccines because they believe it means supporting abortion because of the use of fetal tissue. A number of Catholic leaders and other anti-abortion activists urge opposition to COVID-19 vaccines based on such cell lines. In truth, a good number of Americans don't know whether the vaccines contain aborted fetal cells. In a poll done in late March, 2021, 27% said they did not know whether the Johnson & Johnson vaccine contains aborted fetal cells.[109] This vaccine did use the Per.C6 cell line originating from a fetus aborted in 1985 in its manufacturing process. Thus, say anti-vaxxers, there are sound ethical reasons to reject vaccination.

7. COVID-19 vaccines have an unacceptable risk of side-effects.

Vaccines are commonly singled out in regard to possible side-effects. In fact, the modern resurgence of the anti-vaccine movement began in 1998 with an article published in *The Lancet*, a distinguished medical journal, concerning an alleged link between a vaccine and a serious side-effect: *autism*. The article's imposing title was "Ileal-lymphoid-nodular hyperplasia, non-specific colitis, and pervasive developmental disorder in children."[110] In it was posited a link between autism and children previously having received the measles vaccine. Since that time a chief criticism of vaccines by anti-vaxxers has been the occurrence of serious side-effects.

With respect specifically to COVID-19 vaccines, anti-vaxxers like to appeal to VAERS (the Vaccine Adverse Event Reporting System). This publicly accessible system provides a U.S. database for reporting side-effects. Reports can be submitted by anyone. As of early September, 2021, more than a half-million side-effects associated with the COVID-19 vaccines had been reported. Anti-vaxxers argue that this number, which almost certainly is significantly underreported, demonstrates how very frequently side-effects occur. To them it seems sensible to reject that risk and instead practice healthy living habits instead.

---

[109] Ipsos/Axios Poll (March 19-22).
[110] Wakefield et al., "Ileal-lymphoid-nodular hyperplasia."

Most of the attention in the media has been with respect to a small set of side-effects. We shall have to content ourselves here with four. Some of these have been more associated with certain vaccines (e.g., stroke with Pfizer's and AstraZeneca; genetic alteration with Pfizer's and Moderna's). With specific respect to COVID-19 vaccines, it has been claimed that the following side-effects can result from vaccination (listed in alphabetical order):

- immune system dysfunction;
- infertility;
- rewritten genetic code;
- stroke.

These are hardly minor side-effects and merit attention.

7.1 Immune System Dysfunction

Although COVID-19 vaccines are intended to boost an effective immune system response to the SARS-CoV-2 virus, anti-vaxxers claim they actually create *immune system dysfunction*. Osteopathic doctor Sherri Tenpenny, who identifies herself as an anti-vaxxer, claims in a Bitchute video that COVID-19 vaccines will kill some people outright and that many more will be sickened by them with various autoimmune diseases within 42 days to a year after vaccination.[111] In an article she posted to the website *Vaxxter*,[112] Tenpenny alleges coronavirus vaccines may reduce the viral load in a person's upper respiratory tract (their intended purpose), but also cause "a serious, anti-body-enhanced injury in the lungs." She further maintains this is only part of the problem. The vaccines, she says, not only don't stop infection, they actually enhance it by making it easier for the virus to invade cells.[113]

7.2 Infertility

With respect to *infertility*, we meet again with Mike Yeadon, former Pfizer researcher, who garnered much attention as co-author (with Wolfgang Wodarg) of a petition to Europe's medicines regulator to halt COVID-19 vaccine trials based on the assertion they could cause infertility in women. In the petition it was speculated that mRNA vaccines like Pfizer's might prompt the body's immune system to attack syncytin-1 (a glycoprotein used by the body in the growth and attachment of the placenta).[114] This petition has been widely cited in social media by anti-vaccine proponents.

Robert F. Kennedy, Jr., in a Facebook post claims, "With no data showing COVID vaccines are safe for pregnant women, and despite reports of miscarriages among women who have received the experimental Pfizer and

---

[111] Funke, "COVID-19 Vaccine."
[112] Vaxxter proclaims it delivers "Scientific Articles Exposing Vaccine Myths and Pharma Foibles." The site offers many products—at a cost. (It is a .com website.)
[113] Tenpenny, "Coronavirus Pt. 6." Her support is an article by Liu et al., "Anti-Spike IgG."
[114] Yeadon and Wodarg, "Petition/Motion."

Moderna vaccines, Fauci and other health officials advise pregnant women to get the vaccine."115

Among those who have taken this claim up is Matthew 'Mat' Staver, a lawyer and former Seventh Day Adventist minister (he now identifies as Southern Baptist), who on "Cross Talk"—a program of the Voice of Christian Youth America—extended the general claim to one that pregnant women are having miscarriages simply by being near vaccinated people.116

7.3 Rewritten Genetic Code and Other Bodily Damage

Using his platform on Facebook, influential anti-vax osteopathic doctor Joseph Mercola has claimed the COVID-19 vaccine will "alter your genetic coding, turning you into a viral protein factory that has no off-switch."117 This claim has been echoed by psychiatrist Andrew Kaufman, most publicly through videos. In a YouTube interview, Kaufman argues that a COVID-19 vaccine might use modern technology to generate an electric current to create miniscule holes by which the vaccine would insert genetic proteins to alter human DNA. The effect of this, he warns, is to create "genetically modified organisms."118

The idea that a COVID-19 vaccine can alter the vaccinated person's DNA—rewrite their genetic code—has its genesis in the fact that the Pfizer and Moderna vaccines are mRNA vaccines. Among the alleged results of mRNA vaccines according to some anti-vaxxers is that they create a human 'hybrid,' which some religious anti-vaxxers argue converts a vaccinated person from someone made in the image of God to someone made in the image of Satan, or bearing the mark of the Beast. Others say the mRNA is used to introduce nanotechnology, which rewrites the human genetic code rendering the person no longer human.119

One of the most prominent anti-vax doctors, osteopath Carrie Madej, produced a video seen hundreds of thousands of times. She places the mRNA vaccines (Pfizer's and Moderna's) within the context of a 'Transhumanism' movement—efforts to transform human beings into something else. She warns these vaccines "are completely experimental," and use technologies "that can change the way we live, who we are, and what we are—and very quickly." In speaking of Moderna, she says its claim to fame is 'modified RNA.' In introducing it into the human body, she argues, side effects over an extended period of time will occur, including possibly increased cancer rates, mutagenesis, and increased autoimmune reactions. The problem, she maintains, is the

---

115 Kennedy, https://www.facebook.com/rfkjr/posts/2877302445929819?__cft__[0]= AZX jk2hL2AwEpWPP7wvhb11QJ3cBpgFKC2ec6m1NoglYRI7J7_a64zMSZQ25SA3mA2bKA_G4-Hhclg_Ru90o1vZcczj_iHEU_Iud5RgcfNr2WQIic570VKfP1qx9rryjHOlFfe-JAj3cvG3DobESv-u0&__tn__=%2CO%2CP-R. Also see CCDH, "The Disinformation Dozen," 14.
116 Porter, "How the Evangelical Christian Right."
117 Quoted in Frenkel, "Most Influential."
118 The YouTube video has been deleted. See Reuters Staff, "False Claim: A COVID-19 Vaccine."
119 For examples and more detailed coverage, see Reuters Fact Check, "Fact Check-mRNA Vaccines."

technology; *transfection*, for instance, produces genetic modification. She warns, "If we become genetically modified, we would not be as healthy." Madej contests the claim by pro-vaccine advocates that the vaccines will not alter our DNA or genome. In fact, she argues, by transfection the vaccine agent will be taken up into our genome permanently, replicating itself and altering us. She further argues that because this is a synthetic agent it can be patented, which raises the specter that some company might own part of our genome. Madej then says another part of the vaccine delivery system is the enzyme *Luciferase*, a bioluminescence that can be used to verify vaccination, but also as a "barcode" or "branding" of the individual person. This, she says, means the vaccinated person has become "like a product." A third element, *hydrogel*, she explains, is government-developed nanotechnology that has the ability to connect with artificial intelligence (AI). This, she continues, means the vaccinated person will be continuously connected to devices gathering information from the individual's body. Moreover, information from outside can be sent into the person. So who is behind this? She indicts the Department of Defense, DARPA (Defense Advanced Research Projects Agency), and the Bill and Melinda Gates Foundation.[120]

7.4 Stroke/Blood Clots

The VAERS database (as of early September, 2021) records 2,219 reports of a 'cerebrovascular accident' (.43% of all reported adverse events). The possibility of *blood clots* or *stroke* (specifically cerebral venous sinus thrombosis (CVST)) being caused by COVID-19 vaccines brought significant media attention in the early days of vaccine administration. This attention generated serious investigation. What was discovered is that after the first dose of a two dose vaccination there was some risk of CVST. Epidemiologist Julia Hippisley-Cox reports that of 10 million people exposed to the AstraZeneca (ChAdOx1 nCoV-19) vaccine, there occurred 7 excess events of CVST in the initial 28 days after vaccination. With respect to AstraZeneca's vaccine—which has not been authorized in the United States—in Europe, of 34 million recipients of the vaccine there were 53 possible cases of blood clots (specifically splanchnic vein thrombosis (SVT)), and at least 169 possible cases of CVST.[121] The raw numbers—53 and 169 or more—are frightening.

But what about the vaccines used in the U.S.? With respect to CNS blood clots (i.e., central nervous system thrombosis), of 54 million recipients of the Pfizer vaccine there were 35 possible cases; of the Moderna vaccine there were 5 possible cases out of 4 million recipients.[122] This, says anti-vaxxers, constitutes yet another serious risk of an adverse health event that can be avoided by

---

[120] Although removed by YouTube, the video is still available on other platforms. I viewed it on Vimeo (https://vimeo.com/529524187).
[121] Hippisley-Cox, "Risk," 9–10.
[122] Hippisley-Cox, "Risk," 10.

refusing vaccination. In fact, some add, this possibility demonstrates that the supposed 'cure' may be worse than the actual disease.

8. COVID-19 vaccines actually increase the risk of infection.

A critical claim of the anti-vaccine movement is that vaccines actually *increase* rather than decrease the risk of infection. Anti-vaxxers like to turn to VAERS to support this claim. For example, Robert F. Kennedy, Jr., in a letter to President Biden used VAERS to claim that within the first three months of the vaccine rollout more than 31,000 injuries from COVID-19 vaccines had been recorded, as well as more than 1500 deaths.[123] As of September 8, 2021, COVID-19 itself was reported as an adverse event of vaccination 13,987 times (2.72% of all *reports*, not of all people vaccinated), and death was reported 5,118 times (1% of reports).[124] Once more the numbers are scary.

Anti-vaxxers present this claim in one of two ways: first, that one can contract COVID-19 from the vaccine itself, or second, that vaccinated people get sick more often than unvaccinated ones.

8.1 Contracting COVID-19 after Vaccination and/or from It

People *can* contract COVID-19 after vaccination. This is an indisputable fact acknowledged even by pro-vaccine proponents. A vaccine does *not* guarantee immunity. Vaccinated people who subsequently become infected are said to have experienced 'breakthrough' cases.

But that is not what this anti-vax claim principally has in mind. Rather, as an article posted on *Vaxxter* argues, the critical ingredient used in mRNA vaccines (Pfizer's and Moderna's) *causes the same damage to the body that COVID-19 does*. Citing a paper published in the journal *Circulation Research*, the argument is that the vaccine presents to the body the S (Spike) protein the virus uses to attach to host cells, and it is this S protein that damages human cells, leading to serious health issues. The article asks, "[I]f the main instrument of physical damage is the spike protein, *then why are we injecting people with billions of spike proteins?*"[125]

8.2 Vaccinated People Get Infected More Often than Unvaccinated Ones

The second claim is that vaccinated people get infected more than unvaccinated ones. The so-called 'Bible' of the anti-vax movement is Eleanor McBean's *The Poisoned Needle*. McBean (1905–1989), an advocate of naturopathy, was a prolific anti-vaccine author (who was anti-sugar and anti-doctors, too). Her work has impressed many readers precisely because she presents impressive numbers and quotes from authorities. Thus, in her famous presentation on smallpox she presents a statistical chart showing how in the period 1872–1881, when 96.5% of infants had been vaccinated, some 3,708 deaths due to smallpox occurred, yet a half century later, the period 1932–1941, during which only

---

[123] Weir, "How Robert F. Kennedy Jr. Became the Anti-Vaxxer Icon."
[124] VAERS (https://wonder.cdc.gov/controller/datarequest/D8; jsessionid=26BE0094FD4089DEDA988EC90098), Sept. 8, 2021.
[125] Vaxxter, "New Study Confirms."

39.9% of infants were vaccinated, only 1 death was due to smallpox. She attributes the significant drop not to the cumulative effect of the vaccine but to anti-vax resistance.[126]

McBean bolsters such numbers with quotations from authorities. For instance, her first supportive quote is from the recorder for the city of London who reported that of 155 persons admitted to a hospital with smallpox, 145 of them had been vaccinated. She reports that 93.5% of hospital admissions were vaccinated people.[127]

With specific respect to COVID-19, anti-vaxxers recognize that the time since vaccinations began has been too short to mount much data. Nevertheless, they can point to actual cases, such as the outbreak in Barnstable County, Massachusetts in July, 2021, where 74% of the 469 cases of the disease occurred among the vaccinated.[128] This was reported by the CDC and widely spread among anti-vaxxers through social media.

This claim is often modified to a more modest one: *the vaccinated are just as much a risk to spread COVID-19 as the unvaccinated*. Anti-vaxxers, who often feel singled out for blame for rising infection rates, believe this information needs more attention. They say that if blame is to be apportioned for spreading COVID-19 at least as much should be assigned to the vaccinated.

9. Other, better courses of action exist.

Anti-vaxxers, no less than pro-vaccine advocates, think *prevention* is superior to *treatment*. But anti-vaxxers argue that vaccine prevention is dangerous and that other long-tested preventive means exist. Also they maintain that the pro-vaccine preoccupation with prevention has led to short-changing treatment.

Many anti-vaxxers extol traditional vaccine-alternative favorites like vitamin C or D, and zinc as preventive measures against diseases, including COVID-19. The latter group—vitamins C and D, and zinc—are all familiar to the general public, widely regarded as safe, and thus immediately appealing as alternatives *if* they work.

9.1 Prevention Strategy: Vitamins and Zinc

Vitamins and supplements like zinc are often recommended by health professionals—one area where anti-vaxxers and medical professionals actually might agree. *Vitamin C* (ascorbic acid) is perhaps the most popular vitamin people take as a supplement to their diet. It is considered an essential nutrient and is important to the body in many ways, including boosting the immune system. *Vitamin D* is probably most famous for its role in bone health, but is also thought to benefit the immune system by increasing antimicrobial peptides the body uses as a natural antibiotic and to be valuable in protecting against respiratory infections. *Zinc*, too, supports the immune system, as well as

---

[126] McBean, *The Poisoned Needle*, 10.
[127] McBean, *The Poisoned Needle*, 11.
[128] Brown et al., "Outbreak of SARS-CoV-2 Infections." The event was th annual "Bear Week" gathering of gay men at Provincetown, Massachusetts.

boosting metabolic function. Thus all three are beneficial when taken in appropriate amounts.

Of the three, vitamin D has the best support for possessing value against COVID-19 infection. Research at the University of Chicago Medicine found that having vitamin D levels above those usually regarded as sufficient (30 ng/ml) may lower the risk of COVID-19 infection, especially for African-Americans. Specifically, levels of 40 ng/ml or greater seemed to decrease risk of infection. Moreover, vitamin D deficiencies (<20 ng/ml) seem associated with greater vulnerability to COVID-19. Yet these results were not found for Whites.[129] Since African-Americans as a group are more reticent about receiving medical care, this information might be especially useful for them.

9.2 Treatment Strategy: Early Treatment

Now let's turn to treatment strategies, or dual prevention/treatment ones. Here the key, say anti-vaxxers, is *early intervention*. This is the theme struck by physician Steven Hatfill. He points to 232 studies reported by late April, 2021, involving 3,706 scientists and 358,764 patients, in which significant improvement (64%) in outcomes followed early treatment. He argues such studies have been ignored because they are not favorable to efforts to authorize emergency use of vaccines. Thus, he says, the withdrawal of effective treatments using hydroxychloroquine is unjustified. Further, he argues, the majority of dead Americans from COVID-19 could have been saved by a different strategy.[130]

Osteopath Craig Wax offers a case study contrasting his own experience with that of his primary physician, both of whom contracted COVID-19. Wax had been taking vitamins and a prophylactic regimen of hydroxychloroquine (200 mg weekly) prior to becoming ill, and after his diagnosis continued the vitamins, added to them, and increased his dosage of hydroxychloroquine to 200 mg/twice a day. He got worse and added prednisone and aspirin to his treatment plan. Because of this early aggressive treatment he recovered quickly and fully. His primary care physician's course was far more troubled, ending with hospitalization and intubation. In his judgment the difference is attributable to early treatment built upon a foundation of a healthy lifestyle.[131]

Alongside early treatment, say anti-vaxxers, is the right to choose one's own care. Jane Orient complains that during the COVID-19 pandemic patients have not been receiving their right to *health care of their own choice*. She further complains that this is accompanied by patients being denied a right to know their choices because what physicians like herself advocate for treatment is being censored in the mainstream media as 'misinformation.' Thus potentially life-saving treatment outside a hospital setting is being denied people.[132]

---

[129] Caldwell, "Study."
[130] Hatfill, "The Intentional Destruction of the National Pandemic Plan."
[131] Wax, "The Critical Role of Early Home Treatment."
[132] Orient, "COVID vs the Oath of Hippocrates."

9.3 Prevention/Treatment Strategy: Hydroxychloroquine

More weighty than the use of vitamin supplements is the use of certain powerful drugs. *Hydroxychloroquine* (HCQ), which has been around for over a century, is best known as a drug to treat malaria, but it also has been used successfully with other diseases. Because it has anti-inflammatory properties it can throttle down an over-active immune response. Such factors suggest that hydroxychloroquine might be an effective way to treat COVID-19.[133] In fact, on March 28, 2020, the FDA authorized emergency use of HCQ to treat COVID-19 patients in hospitals. This emergency use was later rescinded.

Voices in the federal government sympathetic to the anti-vaccine movement, including GOP Senators Ron Johnson (Wisconsin), Mike Lee (Utah), and Ted Cruz (Texas), have questioned the FDA decision to rescind authorization. They invited physician Jane Orient of the AAPS to testify before a Senate committee. Consistent with her organization's position, Orient made the following points:[134]

1. The FDA has no authority to regulate the practice of medicine, including the right of individual doctors to prescribe a drug like hydroxychloroquine (HCQ) for so-called 'off-label' use.
2. HCQ is safe.
3. The FDA's emergency authorization restricted use of hydroxychloroquine to those least likely to benefit from it while making it unavailable to those who might best benefit from it.
4. The FDA decision relied on an evidence-based standard the AAPS rejects.
5. Published studies in prestigious journals are suspect.
6. The existing 'standard of care' is a top-down, authority-based model that constitutes "therapeutic nihilism."

But while hydroxychloroquine is still being advocated, to a certain extent it is 'yesterday's news' even for anti-vaxxers. Two other drugs have been receiving more attention.

9.4 Treatment Strategy: Remdesivir

A much touted medication is *remdesivir*, an anti-viral usually administered by injection. It operates by blocking a virus from replicating itself, specifically by interfering with a key enzyme the virus requires for replication. In a double-blind, randomized, placebo-controlled study involving 1,062 hospitalized COVID-19 patients, remdesivir was a superior treatment compared to placebo. A 10 day course of the drug had shorter recovery times and might also have prevented progression to more severe respiratory disease.[135]

---

[133] Sinha and Balayla, "Hydroxychloroquine and COVID-19."
[134] Orient, "Testimony."
[135] Beigel et al., "Remdesivir for the Treatment of COVID-19."

A study reported in the journal *Cell Research* found that remdesivir in a laboratory setting (*in vitro*) inhibited SARS-CoV-2. But it is not a preventive drug; their research showed it functioned at a stage *after* virus entry.[136] The National Institutes of Health, citing this study, noted in the Spring of 2021 that it was approved by the FDA (in May, 2020) for both hospitalized adult and pediatric (age 12 and above) patients to treat COVID-19.

But anti-vaxxers argue this emergency authorization is too restrictive. They contend the drug, given its proven safety and efficacy, should be authorized for much wider use. Restricting its use to severely affected hospital patients seems to them a restriction of their rights to receive the medical care they might desire.

9.5 Prevention/Treatment Strategy: Ivermectin

More recently, *ivermectin* has been championed as a vaccine alternative. Unlike remdesivir and hydroxychloroquine, ivermectin is being praised by anti-vaxxers as a *preventative* medication and, if needed, a treatment after infection. Antibacterial in nature, ivermectin is commonly employed as a preventative medication against heartworm is some small animal species. It is also FDA approved for human use to treat certain parasitic worms and as a topical treatment for external parasites like head lice.[137]

The possible preventative benefit was discovered accidentally in British long-term care facilities when an outbreak of scabies was treated with oral ivermectin (400 or 200 µg kg^{-1}). No resident thus treated developed severe COVID-19 or died.[138]

The use of ivermectin as a treatment strategy has garnered much more attention. For example, a preprint scientific paper much touted by anti-vaxxers claims to demonstrate that ivermectin can reduce COVID-19 deaths by 90% or more.[139] Conservative politicians like Republican Senator Ron Johnson (Wisconsin) and Fox News Host Laura Ingraham have publicized and endorsed its use for COVID-19. Ingraham has added the idea that drugs like ivermectin are being "suppressed" to protect the approved vaccines.[140]

Some studies suggest it might be effective against viruses because in laboratory settings it has been seen to interfere with the process viruses use to enhance infection and inhibit an immune response. With specific respect to the COVID-19 virus, ivermectin might interfere with the spike protein it uses to attach to human cells (the same target of the approved vaccines).

A study published in the *American Journal of Therapeutics* in 2021 examined the repurposing of ivermectin for use against COVID-19. In April of 2021 the researchers examined bibliographic databases and conducted a meta-analysis that yielded 24 randomized controlled studies (with a total of 3406 participants). A meta-analysis of 15 trial studies (2,438 participants) found that use of

---

[136] Wang et al., "Remdesivir and Chlooquine."
[137] FDA, "FAQ: COVID-19 and Ivermectin."
[138] Bernigaud et al., "Oral Ivermectin."
[139] Elgazzar, et al. "Efficacy and Safety of Ivermectin."
[140] Blake, "How the Right's Ivermectin Conspiracy."

ivermectin compared against its nonuse resulted in an average reduction of death by 62% for hospitalized patients. To put this in a different perspective, this is a reduction of mortality rate from 7.8% to 2.3%. The researchers concluded, "The findings indicate with moderate certainty that ivermectin treatment in COVID-19 provides a significant survival benefit."[141]

As of the beginning of September, 2021, some 54 Facebook groups with about 75,000 members were promoting the use and/or sale of ivermectin. They include such groups as Ivermectin MD Team (over 27,000 members), and Ivermectin Covid-19 Testimonials (over 4,000 members).[142] As anti-vaxxers have strongly established, social media is a highly effective means to communicate to one another and to reach the wider public.

10. *COVID-19 will decline without the need for vaccines, etc.*

The last argument we will consider is actually two-fold: that coping with infectious diseases not require any human intervention because they decline naturally, and that natural immunity is better anyway. Let's look at each part.

10.1 Diseases Decline on Their Own.

The first contention is that diseases decline on their own without the need for vaccines. After a disease spreads widely enough to infect many people it eventually mutates into a variant that *only* makes people sick without also killing them. Even if it doesn't entirely disappear, it becomes nothing more than one more nuisance to generally healthy people. In fact, just such a thing happened with the 1918 influenza, which was one of the great killers of history. It eventually mutated into less deadly strains. Although it is too early to say this must happen with COVID-19 it is entirely logical to think that it will. Thus, say some anti-vaxxers, a wait-and-see attitude is not irrational because the odds are that the disease will soon mutate into an even less serious one. This, combined with the first argument discussed in this chapter, makes for a combined claim that not only is COVID-19 not that serious, but it will become even milder with time.

10.2 Natural Immunity Is Superior to Vaccine Immunity

The second contention is that natural immunity (i.e., immunity through contracting the disease) is superior to the 'artificial' immunity of vaccination. The basic claim is often framed much like how it is offered by Shiva Ayyadurai, who has a Ph.D. (engineering) and is a conservative Republican politician. In a video posted to YouTube he argues that *viruses neither harm nor kill, because the human body knows how to take care of itself.* The real danger comes from organizations like the CDC and National Institutes of Health who are "criminal" in not acknowledging such things.[143]

---

[141] Bryant et al., "Ivermectin for Prevention and Treatment." The quote is from p. e452.
[142] Martiny and Gogarty, "Facebook Groups."
[143] Butler, "Anti-Vaxxers."

Let us turn elsewhere. The anti-vax website DangersofVaccines.com, operated by health coach Landee Martin and network marketing professional Walt Shilinsky, presents arguments for 'natural immunity versus artificial immunity.' Their presentation is headed by a claim from pediatrician Paul Thomas that of his more than 13,000 child patients the unvaccinated ones are "by far the healthiest." The main article makes the following arguments: first, the risk of getting sick is exaggerated; second, the vaccine itself can give one the disease (and increase the chance of reinfection); third, the benefit of getting sick is having the immune system strengthened (and otherwise the risk is increased of degenerative diseases later in life); and, finally, people should have the right to choose natural medicine (which is said to be 'under siege').[144]

Anti-vax doctors belonging to a small, recently formed group named 'America's Frontline Doctors' champion the idea of natural immunity. Member Daniel Erickson, with his partner Artin Massih, offered an interview in which they started with well-known and accepted medical facts—that the immune system is built by exposure. They then proceeded to argue that avoiding diseases (like when people shelter away from COVID-19) *decreases* the immune system with the result that when exposure does occur diseases spike.[145]

But perhaps the most sophisticated anti-vax argument for natural immunity runs like this: *viral infections like COVID-19 provoke a natural immune system response; the more people who are infected, the better, because then herd immunity results and the virus is stopped. Most people either do not get sick (asymptomatic), or are not very seriously sick. Infected people who are not very sick spread the disease rapidly among other low-risk people and herd immunity then blocks the virus from infecting those most vulnerable to it.*[146]

This is the line of reasoning espoused by physician Scott Atlas, a radiologist who serves as a Senior Fellow in Health Care policy for the conservative Hoover Institution. He articulated this argument in response to the shutdown that happened early in the pandemic. What he said has been shared over a million times. He builds his argument on what he terms five facts: most people have a low risk of dying from COVID-19; protecting older, at-risk people prevents hospital overcrowding; herd immunity is prevented by shutdown policies, thus prolonging the problem; other people with medical issues are dying because of over-reaction by hospitals; and those that are most at risk can be best protected by policies targeting just them.[147]

Finally, anti-vaxxers appeal to the report issued by the National Vaccine Information Center (NVIC), citing three scientific studies, that, "Research on natural immunity from SARS-CoV-2 infection varies and suggests that durable immunity to the virus lasts for at least eight months and may be life-long."[148]

---

[144] "Natural Immunity vs. Artificial Immunity."
[145] See their comments at Brownstein, "COVID-19 Lockdowns."
[146] See Atlas, "The Data Is In."
[147] Atlas, "The Data Is In."
[148] National Vaccine Information Center, "SARS CoV-2 Virus and COVID-19 Vaccine Information."

Chapter 6
The Pro-Vaccine Rebuttal

Let's start by refreshing our memory of the objections posed in the previous chapter:

1. COVID-19 is not that dangerous.
2. COVID-19 vaccines were developed too fast and are dangerous.
3. COVID-19 vaccines were developed by Big Pharma, which only cares about profit.
4. COVID-19 vaccines are backed by Big Government, which threatens the sanctity of individual decisions about personal medical care.
5. COVID-19 vaccines include dangerous ingredients (e.g., preservative ethylmercury or aluminum salts).
6. COVID-19 vaccines include morally objectionable ingredients (i.e., aborted fetal cells).
7. COVID-19 vaccines aren't safe; they carry an unacceptable risk of undesirable side-effects.
8. COVID-19 vaccines actually increase risk of infection.
9. Other, better courses of action exist (e.g., hydroxychloroquine, ivermectin, vitamin C or D, zinc).
10. COVID-19 will decline without the need for vaccines; natural 'herd immunity' is superior to vaccine immunity.

In what follows I will try to briefly recapitulate the anti-vaccine argument, but the previous chapter always can be reviewed as needed.

1. COVID-19 is not that dangerous.
While the argument that COVID-19 is not that dangerous was a prominent one early in the pandemic, it is not heard as often now, though it remains popular among certain ant-vax scientists and physicians. Remember Mike Yeadon, who proclaimed the COVID-19 virus to be less of a threat to those under age 70 than the seasonal flu? Recall his view that a prediction of some 40,000 deaths in the UK was a sound one?

That prediction, as of early September, 2021—less than a full year later—thus far has missed the mark by more than 90,000 deaths, as 133,000 people (and counting) have perished from COVID-19 in the United Kingdom.[149]

As to his complaint against the popular Polymerase Chain Reaction (PCR) test, one obvious way to examine his claim is to see if that test is confirmed by any others. Fortunately, there are different types of tests, such as viral tests (antigen and nucleic acid amplification tests (NAAT)), and serology (or antibody) tests, and these various tests all continue to find evidence of COVID-19 viral loads in symptomatic and asymptomatic people, young and old, those

---
[149] Gov.UK, "Deaths in United Kingdom" (Sept. 7, 2021).

with and those without pre-existing or underlying health conditions. In the judgment of the majority of scientific experts, COVID-19 is not only real and distinguishable, but widespread and deadly.

But is it any worse than seasonal influenza? Yeadon and many other anti-vaxxers have argued it is not. Yet as of May 11, 2020, the established mortality rate of COVID-19 worldwide was 6.8%, compared to less than 1% for influenza.[150] As pointed out in a previous chapter, compared to influenza, which claims up to 650,000 deaths a year worldwide, COVID-19 is estimated to be *possibly 10 times or more deadly*.[151]

Yet as we saw in the last chapter, German researchers Reiss and Bhakdi dispute that calculation. They find COVID-19 death rates comparable to the seasonal flu, offering an estimate of 0.4%. As they admit, their estimate includes as wide a range of cases as possible, embracing even the asymptomatic ones. Yet *this remains two to four times greater than their own reporting of fatality rates* for an average seasonal flu in Germany (0.1% to 0.2%).[152] More concerning, though, is that the rate comparison pales beside the actual number of dead. A seasonal flu is not a pandemic, and the sheer volume of deaths, even if as small a ratio as 0.4% still adds up to a lot of dead folk given the scale of the infection rate in the general population. That is a key reality anti-vax arguments conveniently ignore. It reminds me of a college I was once employed at where each year the President allocated salary increases as a set percentage for everyone across the board, touting this as completely fair. But it doesn't take a genius to figure out that 5% of $100,000 is substantially greater than 5% of $35,000. The rich get richer. Similarly, the high spread of infection—not the mortality rate—is what qualifies COVID-19 as a pandemic. *Even if influenza and COVID-19 mortality rates were exactly the same, the sheer volume of COVID-19 cases makes it a more serious public health issue.*

The more basic concern of the authors is set forward in their Preface, where they complain of "civil right restrictions" not seen since the end of World War II and the "collapse of social life and the economy."[153] Like other anti-vaxxers, their real quarrel is with entities like government and the media.[154] Plus, as is also true of anti-vaxxers, they embrace a culture of individualism where individual risk is considered far more important than public risk.

One might expect the mortality rate in the United States might be better than that worldwide, but the data from that same period in 2020 discussed above suggests how serious COVID-19 is for Americans. An article published by physician researchers Jeremy Samuel Faust and Carlos Del Rio reported that the number of American deaths from COVID-19 just in the week ending April 21, 2020, was from 9.5–44.1 fold greater than the *peak* week of deaths from

---

[150] Buonaguro, Ascierto, and Morse, "Covid-19."
[151] Maragakis, "Covid-19 vs. the Flu."
[152] Reiss and Bhakdi, *Corona False Alarm?* See chapter 2.
[153] Reiss and Bhakdi, *Corona False Alarm?* Preface.
[154] Reiss and Bhakdi, *Corona False Alarm?* See ch. 8.

influenza over the preceding seven years (with an average of 20.5 fold greater)—and, if anything, COVID-19 deaths were undercounted.[155]

But what about Joseph Lapado's call for "rational decision making," a cost/benefit analysis including many other things in life that people value and not merely COVID-19? He does not indicate even in broad strokes how such an analysis might be conducted. But he does offer a conviction that probably serves as his yardstick: *liberty*. In his estimation, 'Covid mania' has led to curtailment of personal liberty. Thus the calculus for deciding something like vaccination—or any other measure related to COVID-19—must be individually made. In short, cultural ideology must triumph.

2. *COVID-19 vaccines were developed too quickly and are dangerous.*

As we saw last chapter, this argument advances two claims: first that the vaccines for COVID-19 were developed too quickly, and second, that they are dangerous. These two claims are sometimes joined in a modified and more modest claim that at best the vaccines' effects are not yet adequately known. Let us examine each claim in turn.

2.1 The Vaccines Were Developed too Quickly

First, anti-vaxxers commonly point out that development of vaccines requires *decades* more often than *years* and thus the development of any vaccine in a year or less looks suspiciously rushed. If history is made the measuring stick that sounds pretty convincing. On the other hand, it took all of human history to produce the first manned flight (1783, in a balloon). Yet from then on, because of technological advances, manned flight advanced rapidly: 1891, in a glider; 1903, the Wright brothers' airplane; and 1969, landing on the moon. Simply reckoning how long things took place in the past ignores the context of changes in technology (i.e., practical scientific advancement).

With respect to COVID-19 vaccines, in addition to some 90 years of previous study of coronaviruses, the technological development of being able to rapidly decode a virus' genome has been a critical factor in the speed by which vaccines can be developed.[156] Another technological advance of great importance has been the ability to rapidly share data among scientists—something far harder just a half-century ago.[157] These points, and others (such as economic support), were covered in earlier chapters.

As a historical note, it might be mentioned that while it is frequently said that the previous record for a vaccine was four years, the urgency of the Ebola virus crisis produced in less than a year a dozen clinical trials that moved from a 'first in man' dosing study to phase III efficacy trials.[158] The keys then were much the same as now: advanced technology and widespread collaboration.

---

[155] Faust and Del Rio, "Assessment of Deaths," 1045.
[156] Ellis, et al., "Decoding Covid-19."
[157] Le Guillo, "Covid-19."
[158] Petousis-Harris, "Assessing the Safety of COVID-19 Vaccines," 1205.

We saw eminent immunologist (and vaccine developer) Ian Frazer quoted as commenting on the problems associated with developing a coronavirus vaccine. The interview which drew anti-vaxxer attention was offered early in the pandemic, as efforts were underway to develop a vaccine for COVID-19.[159] He was speaking of efforts to develop a SARS vaccine (not specifically a COVID-19 one) and observed that it had caused inflammation in the lungs of animals to which it had been administered. Obviously, he pointed out, if you produce a vaccine with the same mortality rate as the disease, it isn't a very rational choice to vaccinate. But he also observed that those developing COVID-19 vaccines are aware of this history. His prediction was that vaccine developers would focus on using a part of the virus attached to a chemical to induce an immunological response. He noted, "That [vaccine type] has been successful in animal models for coronaviruses in the past." The key in a narrow, targeted approach, he emphasized, is selecting the right agent to provoke the immune response. Frazer was *not* declaring the development of a vaccine impossible, or even as taking overly long (he thought one could be developed within 18 months).[160]

2.2 The Vaccines Are Dangerous

The general argument that vaccines are unsafe/dangerous—and even the Supreme Court says so!—is, of course, an appeal to authority. But let's examine it. Justice Scalia's opinion for the court's majority begins by acknowledging the tremendous success of vaccines and repeating the CDC judgment that they collectively constituted one of the greatest 20th century achievements of public health. Scalia went on to observe that their very success left a public less worried about the diseases than the vaccines. In this respect the 1980s saw an increase in vaccine-related lawsuits. This in turn had an adverse effect on vaccine makers, some of whom decided to discontinue their efforts. To address this situation Congress in 1986 created the National Childhood Vaccine Injury Act (NCVIA). The Act was designed to protect vaccine makers against spurious lawsuits while also protecting those with legitimate claims. In the case before the Court—Bruesewitz v. Wyeth (2011)—the vaccine in question was for DPT. Scalia parses the language of the NCVIA. He then notes that the FDA has never specified the criteria by which a vaccine is safe and effective for its intended use. However, he immediately adds, "And the decision is surely not an easy one. Drug manufacturers often could trade a little less efficacy for a little more safety, but the safest design is not always the best one. Striking the right balance between safety and efficacy is especially difficult with respect to vaccines, which affect public as well as individual health."[161]

The anti-vax argument focuses on the words 'unavoidable, adverse side effects.' When Scalia uses these words it is in reference to the NCVIA language

---

[159] The full interview can be viewed on YouTube (https://www.youtube.com/watch?v=wKxhaWYkB3I).
[160] In addition to the YouTube interview, see Kahn, "We've Never Made a Successful Vaccine."
[161] Supreme Court of the United States, "Syllabus: Bruesewitz et al. v. Wyeth et al." The quote is from Scalia's opinion, p. 13.

that shields vaccine makers from *all* claims of adverse effects because not all adverse effects rise to the level of harm the anti-vax argument presumes. Slight redness at the site of an injection, for example, is an adverse event, but hardly one worthy of a lawsuit. The term 'unavoidable' simply refers to the possibility of making some other vaccine where a particular adverse event *couldn't* happen. Thinking that way opens legal action to endless speculative actions. The point is that once a vaccine is actually made, some adverse events become possible and because some are possible there will be instances where they are 'unavoidable' in a legal sense.

The *Vaxxter* article very disingenuously suggests in connection with the above case that legally vaccines are dangerous. The word 'dangerous,' however, does not appear in the Supreme Court Syllabus for the case. Instead, it appears in an article by law professor Ellen Wertheimer and her recommendations to courts as to how they might decide cases. In her thinking, vaccines are an example of the type of dangerous products "the utility of which outweighs their dangers and which are available and in use for the good of society generally."[162] When Scalia uses the term 'unsafe,' it is in the context of saying that "whenever the FDA concludes that a vaccine is unsafe, it may revoke the license."[163] That is hardly the same as saying that all vaccines are unsafe or even that any particular one is.

But what about the scientific studies anti-vaxxers point to? These suggest that unvaccinated children are healthier than vaccinated ones with respect to a range of health conditions. The first one cited compared the two groups with respect to developmental delays, asthma, ear infections, and gastrointestinal disorders. This study looked at 4,821 children, all from one or another of three pediatric practices. It was specifically interested in comparing those *unvaccinated in infancy* to those *vaccinated in infancy* (i.e., before the first birthday). The former group consisted of 31% of the whole (633 children); of them, 84% remained unvaccinated, with the remaining 16% receiving one or more vaccination after their first birthday. The strongest association was found for developmental delay (6.6% to 2.4%)—still modest numbers in both groups. The authors note that the chief limitation of their study is that it was a convenience sample and might not represent the general population. Their own recommendation was not to avoid vaccination, but rather conduct further studies to test the reliability of their findings.[164] The other study mentioned in the last chapter on this point was retracted based on "several methodological issues" and conclusions "not supported by strong scientific data."[165]

---

[162] Wertheimer, "Unavoidably Unsafe Products," 189.
[163] Supreme Court of the United States, "Syllabus: Bruesewitz et al. v. Wyeth et al." The quote is from Scalia's opinion, p. 15.
[164] Hooker and Miller, "Analysis of Health Outcomes."
[165] IJERPH Editorial Office, "Retraction" (https://www.mdpi.com/1660-4601/18/15/7754/htm).

The safety of COVID-19 vaccines is typically linked directly to how quickly they were developed and approved, so this matter requires addressing. Despite the speed with which COVID-19 vaccines have appeared, rigorous safety reviews were conducted before emergency use approval—and continue. Anti-vaxxers don't bother to point out that the trials before emergency approval involved larger numbers of participants than what is typical for vaccine trials. Moreover, the FDA asked for a median of two months of follow-up safety data after complete vaccination by trial participants.[166] A meta-analysis of the safety profile of these vaccines published in 2021 found, "the available evidence indicates that eligible COVID-19 vaccines have an acceptable short-term safety profile."[167] And an article in *Nature*, after more than 1.7 *billion* doses of the vaccine had been distributed, found that vaccine safety and efficacy both remained quite high.[168]

3. COVID-19 vaccines are really just about profit.

In the last chapter we heard how prominent anti-vaxxer Robert F. Kennedy, Jr. made the argument that the handling of vaccines undermines capitalism. Like Don Quixote, anti-vaxxers such as Kennedy tilt against the windmills of Big Pharma, thus casting themselves as the heroic little guy fighting for American rights against the callous calculations of greedy pharmaceuticals. And they can certainly find enough ammunition to support charges of corporate greed, intentional misleading of the public, and disregard for adequate patient safety. I doubt many Americans will contest such a general picture. But does it apply specifically to vaccines, and most particularly to COVID-19 vaccines?

Prior to the COVID-19 pandemic the best descriptor for the pharmaceutical involvement with vaccines was "decline." In the last half of the 20th century and on into the 21st century the number of pharmaceutical companies engaging in making vaccines reduced dramatically. Even those still involved reduced their budgets for research. Why?—because the research, development, testing, and manufacture were not cost effective in a shrinking market.[169]

The pandemic brought a rush of new involvement. As we saw in an earlier chapter, scores of vaccines are in some stage of research and/or development. In this rush anti-vaxxers only see greed with Big Pharma's lust for profit overwhelming concerns for safety. But among other problems with this scary picture is that such major pharmaceutical companies as Johnson & Johnson and AstraZeneca declared early on their adopting a not-for-profit approach to pricing the vaccine.[170] And, as was pointed out in an earlier chapter, the estimated benefit relative to the price has been estimated at anywhere from 40–300 times greater. Even if companies are actually making substantial profit there

---

[166] Ellenburg et al., "The Long View."
[167] Wu et al., "Evaluation of the Safety Profile," 12.
[168] Ledford, "Six Months of COVID Vaccines."
[169] Offit, "Why Are Pharmaceutical Companies Gradually Abandoning Vaccines?"
[170] Weintraub, "Pfizer CEO."

exists a strong economic argument that what Americans are getting for their money is a tremendous bargain.

As we have seen before, the real goal of the anti-vax arguments on this point is to shift attention away from science to the cultural issues they are interested in. With respect to profit, they want people talking not about the economic benefits of vaccines but about how large institutional structures like government and business make decisions at the expense of individuals who want to make contrary choices—like profiting from anti-vax alternative products. In fact, as long as we are on the subject of profit motives, we should also consider an aspect of the anti-vaccine movement that its leaders prefer not receive any attention—their own profiteering.

A collaborative study among the Public Data Lab, Digital Methods Initiative, Open Intelligence Lab, and First Draft analyzed the web analytic trackers of 60 anti-vaccination websites. They found such sites use information from trackers not merely to facilitate traffic to their sites, or build community, but also to target and serve advertisements. They also found, by analyzing the interfaces and codes of these sites that they build ways to monetize their content, such as to sell books and direct visitors to buy products (e.g., alternatives to the vaccine).[171]

The Center for Countering Digital Hate (CCDH), a non-profit non-governmental organization (NGO), issued a 72 page report on the anti-vaxxer movement. Based on the information they were able to access, they estimated the worth of this movement to technological enterprises like social media to be more than a billion dollars per year (helping to explain the tolerance of such platforms to misinformation despite cries of 'foul' by experts and the general public). The anti-vax 'industry' itself—the money generated from anti-vax publications and the like—is estimated to be in excess of $35 million a year.[172]

Now in the larger scheme of things this is certainly a modest amount, especially compared to Big Pharma. But it may be worth observing that the concentration of wealth is among a few individuals—people we might term 'professional anti-vaxxers.' In fact, reports CCDH, a mere dozen individuals account for up to 70% of anti-vax content on Facebook, and of that dozen, just *three* are responsible for more than half of that amount.[173]

Let's consider one of these three—Joseph Mercola, who heads a business worth $100 million in 2017. He fits the traditional profile of an anti-vaxxer, being an advocate of alternative and natural medicine, which he is happy to supply through a variety of products. In fact, in response to the pandemic he quickly began offering alternative remedies to COVID-19 through his website *Stop Covid Cold*.[174] No less than Big Pharma, anti-vaxxers like Mercola have a

---

[171] Smith, Bounegru, and Gray, "How Anti-vaccination Websites."
[172] CCDH, *Pandemic Profiteers*, 4.
[173] CCDH, *Pandemic Profiteers*, 5. Cf. CCDH, "The Disinformation Dozen," 7.
[174] CCDH, *Pandemic Profiteers*, 18. The website was discontinued following warnings from the FDA concerning misleading claims. Previously, in 2006, the Federal Trade Commission forced him to

profit interest. It is a self-vested economic interest of anti-vaxxers to sow distrust in the vaccines (which are bad for business).

Just as Big Pharma strategizes market moves, so does the Anti-Vax Industry. CCDH found that in October, 2020, a conference was held at which the anti-vaxxers in attendance decided to push a multipronged attack: minimize the danger of COVID-19, maximize the risks of vaccination (e.g., side effects), and generally subvert health experts and otherwise impede vaccination.[175] On the positive side for their businesses, affiliative marketing schemes increase profit. These schemes work through anti-vaxxers cooperating to promote one another's profiteering as 'affiliates.'[176] This strategy conveys the impression to audiences of a large and diverse group of people sharing a common goal rather than a few homogenous groups or individuals collaborating for mutual profit.

Members of the anti-vax (or quasi anti-vax) physician organization the Association of American Physicians and Surgeons (AAPS) also have been implicated in profit-motivated (or at least vested self-interest) stances. Researcher Kathleen Seidel found instances of articles published in the organization's journal by authors presently involved in litigation when they published articles purporting findings favorable to their case, which they could then cite in court as having appeared in a 'peer-reviewed' journal.[177] Such acts are at best morally dubious and certainly violate scientific and medical ethics. But they have a history in anti-vaxxer circles, as we shall presently see in the case of Dr. Andrew Wakefield.

4. *COVID-19 vaccines are backed by Big Government, which interferes with individual decisions.*

As noted when this argument was introduced, it more than any other strikes to the heart of the matter for anti-vaxxers. We saw Yeadon's worry that governments around the world are colluding to gain control over the lives of the individuals in their nations. Why? He fears it may be for mass depopulation, either via mass deportations, or more worryingly, through mass vaccination to kill people. He thinks eugenicists have accumulated enough power to hold the levers. He invokes the spectres of Hitler, Stalin, and Chairman Mao.[178] No actual evidence is offered, and Yeadon frequently couches his claims as "possibly" true, then argues the mere possibility is justification enough for action. But there is a difference between 'possible' and 'probable,' with only the latter calling for action. After all, it is *possible* the sun will supernova tomorrow,

---

stop making claims that the tanning beds he sold would reduce the risk of cancer (he was made to refund $2.59 million to over 1,300 customers). Mercola continues to post anti-vaccine articles which claim to be 'fact-checked,' though the fact-checking is done by an employee, not an independent researcher. As Frenkel, "The Most Influential Spreader," notes, both Facebook and Twitter have felt compelled to take action against Mercola.

[175] CCDH, *Pandemic Profiteers*, 4.
[176] CCDH, *Pandemic Profiteers*, 14.
[177] Seidel, "Strange Bedfellows."
[178] Delaney, "EXCLUSIVE—Former Pfizer VP."

but it isn't *probable*. The inherent difficulty in refuting conspiracy theories is that because they rely so completely on fear-inspired belief they are immune to such troubling inconveniences as actual facts.

We saw Jane Orient's statement, which provides a convenient example of this kind of argument. Its initial declaration strikes the theme at the heart of the argument: "The Association of American Physicians and Surgeons (AAPS) strongly opposes federal interference in medical decisions, including mandated vaccines."[179] It is a culturally-based objection.

Orient's statement focuses much less on science than on cultural values. Her basic points are, first, that patients have *personal rights*, for example, to informed decisions about medical care or making decisions for their children. Second, justification for mandated vaccines assume a public health risk, but what about the "threat" posed by "forcing" people to assume "government-imposed risks"? She claims that all vaccines are by their nature risky, but her central argument with respect to them is the risk-benefit ratio involved must be decided not by the federal government, but by individual patients and physicians.

This sounds like a highly reasonable argument. Why should the federal government (or state or local) decide for a person what the person can decide for her- or himself? However, consider the logic applied widely. Why can't each individual decide whether to wear a seat belt or pay taxes? Of course, they can; but any choice comes with consequences (including vaccination refusal). But as a collection of individuals, the American public has authorized government at various levels to act in the common interest so as to safeguard the well-being of all people. Just as civil authorities have been granted—by the citizens—the right to make regulations for seat belts and to collect taxes, the government has a right to insist on vaccine mandates.

As an interesting aside, Orient does *not* exclude vaccination as an individual choice. With respect to measles, she advocates as a better public health measure the development of "a more effective, safer vaccine." She also points out that there exists a public right to restrict the movements of "potentially contagious individuals" when the disease presents a "clear and present danger."[180] This is an important recognition that personal freedom has boundaries.

Interestingly, when anti-vaxxers are in high positions in government the role of Big Government does not get the same level of attention. Instead, blame is shifted to some other convenient target. Thus, when physician Joseph Lapado was appointed to become Florida's Surgeon General he blamed public mistrust not on the actions of government, but on *scientists*. As we saw in the last chapter he claims some scientists misrepresent the science on COVID-19 for their own ends, thus inculcating a climate of fear at the center of Public Health. He announced his intention of replacing such fear, and the mistrust behind it, with

---

[179] Orient, "Statement." The statement actually has a *measles* outbreak in mind.
[180] Orient, "Statement."

a positive agenda driven by the cultural ideals favored by anti-vaxxers, namely freedom of choice and natural health alternatives (e.g., better eating (more fruits and vegetables), losing weight, and exercise). Like Orient, he does not expressly reject vaccination; he just regards it as insufficient and a lesser alternative.[181]

5. COVID-19 vaccines include dangerous ingredients.

A common argument made against vaccines is that they contain dangerous ingredients. When such claims are left vague and divorced from context they can promote considerable fear. After all, most Americans presented a list of ingredients for a vaccine are left shaking their heads as to what danger, if any, they might pose simply because they don't know what those ingredients are—even if named.

Let's begin by considering the inactive ingredients—the ones the anti-vax argument makes claims about—for the three approved COVID-19 vaccines in the United States[182]:

Johnson & Johnson	Polysorbate-80; 2-hydroxypropyl-β-cyclodextrin; Citric acid monohydrate; Trisodium citrate dihydrate; Sodium chloride; Ethanol
Moderna	PEG2000DMG: 1,2 -dimyristoyl-rac-glycerol, methoxypolyethylene glycol; 1,2-distearoyl-sn-glycero -3-phosphocholine; Cholesterol; SM-102: heptadecan-9-yl 8-((2-hydroxyethyl) (6-oxo-6-(undecyloxy) hexyl) amino) octanoate; Tromethamine; Tromethamine hydrochloride; Acetic acid; Sodium acetate; Sucrose
Pfizer	2[(polyethylene glycol (PEG))-2000]-N,N-ditetradecylacetamide; 1,2-distearoyl-sn-glycero-3-phosphocholine; Cholesterol; (4-hydroxybutyl) azanediyl)bis(hexane-6,1-diyl)bis(2-hexyldecanoate); Sodium chloride; Monobasic potassium phosphate; Potassium chloride; Dibasic sodium phosphate dihydrate; Sucrose

Obviously, it requires rather specialized knowledge to readily comprehend such lists.

Neither aluminum nor thimerosal appear on the above lists, but the persistence of such claims means both need a closer look.

5.1 Aluminum (Alum)

Let us consider first the ingredient *alum* (aluminum oxy-hydroxide). As observed in the last chapter, it has been used in vaccines since the 1930s as an adjuvant (an ingredient that enhances immune system response). The exact mechanism(s) by which the desired effect is achieved remain uncertain, and gives even scientists pause. Research studies have shown that alum *if* included in COVID-19 vaccines would promote high concentrations of antibodies.[183]

Anti-vaxxers, remember, point out the inherent neurotoxicity of aluminum and remind us that some research finds toxic side-effects related to vaccination

---

[181] See Mower and Wilson, "Florida's Next Surgeon General."
[182] CDC, "Frequently Asked Questions" (Sept. 3, 2021). N.B. The *active* ingredient is the key agent meant for human immune response; that topic is covered in the previous chapter and touched upon in this one. Also see FDA, "Vaccine Information Fact Sheet."
[183] Hotez et al., "COVID-19 Vaccines," 399.

beyond actual allergic reaction. In this regard, a 2015 paper authored by Gherardi et al. in *Frontiers in Neurology* is sometimes referenced. That paper found that aluminum adjuvants have a role in macrophagic myo-fasciitis (MMF) lesions. However, the same study refuted the belief *every* person vaccinated will experience MMF lesions. In the small number of adversely affected folk, MMF has been associated with chronic fatigue and other forms of myalgia (muscle pain). The researchers do *not* call for the elimination of aluminum adjuvants but suggest further investigation to see "why a given individual will appear intolerant to alum-containing vaccines whereas the vast majority of individuals vaccinated with the same vaccines remain healthy."[184] Interestingly, while anti-vaxxers may cite the article, they do not delve into the details, because it is the *appearance* of being backed by science that matters more than the *reality* of what the science says.

Physiologist and well-known science educator Jonathan Berman points out that over 90 years of empirical evidence regarding alum's use in vaccines consistently demonstrates its safety. Moreover, the actual amount included is comparable to that found in infant formulas (and even natural breast milk has some aluminum).[185] Unfortunately, in the interest of fear-mongering to drum up support anti-vaxxers conveniently ignore inconvenient evidence.

However, with respect to COVID-19 vaccines, the concern is moot because those vaccines authorized in the U.S. do *not* contain aluminum.[186] Only 8 vaccines in the U.S. do (anthrax; hepatitis A and B; human papillomavirus (HPV); diphtheria-pertussis-tetanus; Japanese encephalitis; and pneumococcal conjugate vaccines).[187]

5.2 Thimerosal/Ethymercury

First, note that ethylmercury is not the same as mercury. A common ingredient in multidose vaccine vials is *thimerosal*, an ethylmercury-containing preservative. It can prompt minor reactions like redness or swelling at the place where the vaccine is administered by a shot. Even though decades of research showed no significant risks associated with thimerosal, in 1999 an agreement was reached with drug manufacturers to further reduce or eliminate it.[188]

With respect to the ingredients usually mentioned by anti-vaxxers, substances like ethylmercury are typically excluded under U.S. law from inclusion, while both it and aluminum salts, even when present in a vaccine, are

---

[184] Gherardi, et al., "Biopersistence." The quote is from p. 4. Also see Gherardi and Authier, "Macrophagic Myofascitis."
[185] Berman, *Anti-Vaxxers*, 98–100 (quote is from p. 98).
[186] Swenson, "US and EU COVID Vaccines."
[187] Hotez et al. "COVID-19 Vaccines," 400.
[188] CDC, "Thimerosal and Vaccines." It is worth noting that ethylmercury is broken down and discharged from the body significantly more rapidly than the methylmercury ingested through diet; see Children's Hospital of Philadelphia, "Vaccine Ingredients—Thimerosal."

in such small amounts that no adverse links to health ever have been found.[189] As with aluminum, *no* use of thimerosal occurs in COVID-19 vaccines.[190]

6. *COVID-19 vaccines include morally objectionable ingredients.*

Some people fear that supporting COVID-19 vaccines means supporting abortion because of an alleged use of fetal tissue. It is true that *some* proposed COVID-19 vaccines follow from research conducted using fetal tissue cell lines. In fact, at least five candidate COVID-19 vaccines use one of two human fetal cell lines. Four of these use such cells to generate nonreplicating adenoviruses to transport select genes of SARS-CoV-2 to prompt cells to produce proteins that will elicit an immune system response. The fifth—a protein subunit vaccine—uses the fetal cell material to produce the virus' S protein and thereby trigger an immune response.[191] But none of these five are approved in the United States, and the mRNA vaccines that are approved do *not* use fetal cell lines. The Johnson & Johnson vaccine did employ cells derived from the Per.C6 line of a fetus aborted in 1985, but that is not the same as actual fetal tissue. The fetal cell lines have been laboratory grown for decades now. Those used during the production process for the J&J vaccine are filtered out and are not present in the final vaccine product.[192]

Because this is a matter that especially evokes a strong emotional response in conservative Christians and Roman Catholics, perhaps we can turn to a leader in such circles for comment. Prominent Evangelical personality James Dobson tells his followers that vaccine makers AstraZeneca and Johnson & Johnson use cell lines which had an origin more than three decades ago in aborted fetal cells. He notes that though the current cell lines used by these pharmaceutical companies contain no human fetal cells, the history remains concerning to him. He commends companies like Pfizer and Moderna, whose vaccines do not use such cell lines.[193] The Roman Catholic Church similarly has set out such thinking, preferring the mRNA vaccines, but also specifying that even the J&J vaccine is morally acceptable if it is the only one readily available.

The bottom line for most people on this matter is that *none* of the U.S. approved COVID-19 vaccines contain aborted fetal cells. But the bottom line for the anti-vax movement is that this line of attack is a proven way to shift attention away from science to cultural considerations, in this case religious beliefs. In this way they hope to capitalize on cultural ties to gain support in their anti-vaccine stance.

For vaccine-hesitant people preventive medicine doctor Richard Zimmerman offers four approaches to resolve ethical concerns. First, he points out that be morally culpable for an act requires a number of factors. For example,

---

[189] See Camargo Jr., "Here We Go Again," 3–4.
[190] Reuter's Staff: Fact Check: COVID-19 Vaccines."
[191] Wadman, "Abortion Opponents Oppose COVID-19 Vaccines' Use of Fetal Cells."
[192] See Schimelpfening, "No, Fetal Tissue Wasn't Used."
[193] Dobson, "The Unprecedented Production."

driving on a road in the Old South built by slaves is unlikely to cause today's driver to feel moral complicity in slavery, because the driver is far separated by both time and intent from the original act, which he or she may not even be aware of having happened. Similarly, vaccines today are remote in time and intent from the abortion event that produced the cell lines, which themselves have multiplied so many times none of the original aborted material any longer exists. Second, an element of altruism—a willingness to act to protect or help others—ethically supersedes morally questionable cell lines. In other words, a higher good provides for a moral choice to be vaccinated to protect others. Third, both religious leaders and sacred texts offer support for disease prevention, which applies to vaccination. Finally, mRNA vaccines were not designed, developed, or produced with any connection to fetal cell lines, so they can pose no ethical objections on such grounds.[194]

7. COVID-19 vaccines have an unacceptable risk of side-effects.

Perhaps no medical treatment, including drugs as common as aspirin, can guarantee a zero possibility of side-effects. Yet vaccines are commonly singled out in this regard. In fact, the modern resurgence of the anti-vaccine movement began in 1998 with an article published in *The Lancet*, a distinguished medical journal. The article's imposing title was "Ileal-lymphoid-nodular hyperplasia, non-specific colitis, and pervasive developmental disorder in children."[195] In it was posited a link between autism and children previously having received the measles vaccine. This immediately generated a concerted scientific effort to test the paper's claims.[196]

Subsequent researchers could not replicate the results and some years later, journalist Brian Deer began a proverbial 'following the money' investigation of Andrew Wakefield, the lead author. Deer uncovered that Wakefield had been paid £435,000 ($609,000 at present exchange rates) by a personal injury lawyer desiring to build a case against vaccine manufacturers. Substantive fraud was established as existing in the paper's claims. In 2010, *The Lancet* published a retraction of the article.[197]

Meanwhile, Wakefield was willing to play both sides of the fence; he also had filed a patent for a rival vaccine to the existing one, hoping to persuade investors to back his own measles shot.[198] When that venture failed, and he lost his license to practice medicine because of the numerous ethical violations linked to the 'research' purported to underlie *The Lancet* article, he willingly took up the mantle of martyr to the anti-vax cause and subsequently earned, by the middle of 2020, nearly a half-million American dollars for his efforts.

---

[194] Zimmerman, "Helping Patients with Ethical Concerns."
[195] Wakefield et al., "Ileal-lymphoid-nodular hyperplasia."
[196] For a careful review of Wakefield's paper, see Berman, *Anti-Vaxxers*, 69–86.
[197] "Retraction."
[198] CCDH, *Pandemic Profiteers*, 20.

So there is a distasteful history behind this particular argument. But that does not mean we should not investigate current claims related to COVID-19 vaccines. Moreover, anti-vaxxers are not particularly interested in the normal, expected reactions that signal the vaccine is working (e.g., redness, pain or swelling at the injection site, headache, chills, or fever). Nor are they typically focused on allergic reactions like anaphylaxis, which the CDC occurs in two to five people per million.[199]

As we saw, with respect to COVID-19 vaccines, anti-vaxxers like to appeal to VAERS (the Vaccine Adverse Event Reporting System). This publicly accessible system provides a U.S. database for reporting alleged side-effects. Reports can be submitted by anyone.

I searched the VAERS database for reports concerning symptoms following COVID-19 vaccination. I did not employ any filters. The system returned 514,270 reports. Some of the more commonly reported or publicized adverse events were[200]:

abdominal pain (11,137)

abnormal sensations (17,779)
aguesia (loss of taste—2,959) or dysgeusia (altered taste—3,512)
anosmia (loss of smell—2,256)
arthralgia (joint pain—30,986)
asthenia (tiredness—20,597), fatigue (79,690), lethargy (5256)
back pain (9,822)
burning sensation (4,972)
CVAs (strokes—2,219)
chest pain/discomfort (24,908)

chills (73,123)/hot flush (2,986)
confusional state (3,956)
cough (13,026)
decreased appetite (8,619)
diarrhoea (17,386)

discomfort (3,005) or pain (69,924; in extremity, 50,878)
dizziness (57,412)

dysphagia (swallowing—3,887)
dyspnoea (labored breathing—27,488)

ear pain or discomfort (4,537) /eye pain or swelling (4,539)
elevated temperature (6,099) or fever (80,814)
epistaxsis (nosebleed—1,721)

falling (5,173) or gait disturbance (4,984)
feeling hot (9,452)/cold (5,482)
flushing (7,681)

headache (97,994)
heavy menstrual flow (2,446)
hypertension (4,752)
hypoaesthesia (diminished sense of touch—16,138)
influenza-like illness (7,022)
insomnia (5,448)
malaise (14,900)
migraine (6,430)
muscle spasms (5,671) or myalgia (muscle pain—32,141)
nausea (56,358)

neck (8,021) or
throat pain (8,055)
pruritis (itchy skin—26,700)
swelling of lymph nodes (14,929)

Repeating a number as more than a half-million alone is an effective way for an anti-vaxxer to declare the vaccines are 'dangerous.' The impression is

---

[199] CDC, "Selected Adverse Events."
[200] VAERS (https://wonder.cdc.gov/controller/datarequest/D8;jsessionid=26BE0094FD4089DEDA988EC90098), Sept. 8, 2021.

reinforced by looking at the symptoms mentioned (and not all of even common symptoms were listed). No one I know wants to experience *any* of those things.

But let's try to put these numbers in perspective. First, ask 'Is the cure worse than the disease?' When one considers the symptoms associated with COVID-19 it becomes more understandable why so many have chosen vaccination. COVID-19 symptoms include[201]:

- confusion (i.e., newly experienced);
- congestion and/or runny nose;
- cough;
- diarrhea;
- fatigue;
- fever or chills;
- headache;
- loss of sense of taste or smell;
- muscle aches or pains;
- nausea and/or vomiting;
- persistent pain or pressure in the chest;
- shortness of breath or difficulty breathing;
- skin discoloration;
- sleep disturbance (i.e., difficulty staying awake or inability to wake);
- sore throat.

But COVID-19 also can lead to hospitalization and death.

Second, all the reported numbers must be seen in a particular context. The reports of dizziness, for instance, account for 11% of all *reported* adverse events, but would be significantly lower for all *actual* vaccinations.

Third, some reported events may be related to the *experience* of getting vaccinated (the setting, or fear of needles), rather than the actual vaccine. For instance (using reported events not listed above), are events like anxiety (6,189), a cold sweat (3,387), dry mouth (1,765), abnormal sweating (18,067), an elevated heart rate (10,182), palpitations (9,998), or even loss of consciousness (e.g., from fainting—8,577), more likely attributable to the vaccine experience or the vaccine itself? It may not be an either/or matter, but certainly many cases of such adverse events are more likely due to the experience of vaccination rather the vaccine itself.

Finally, fourth, the VAERS system is malleable; it can be manipulated. Some evidence suggests that it is used by anti-vax advocates to spread misinformation by selective reporting of numbers taken out of context, using select reports as 'testimonials' of vaccine danger, or even by creating false reports.[202] A notable example of the kind of thing that can happen is when one person is

---

[201] See CDC, "Symptoms of COVID-19."

[202] For more on this matter, see Brumfiel, "Anti-Vaccine Activists."

alleged to have filed a report saying the vaccine turned him into the incredible Hulk.

The key in understanding the database is that these are *alleged* side-effects. The reports, remember, are *unvetted*, meaning they are unreliable sources. The VAERS database itself explicitly declares, "The reports may contain information that is incomplete, inaccurate, coincidental, or unverifiable. In large part, reports to VAERS are voluntary, which means they are subject to biases."[203]

Now most of the side-effects listed above—even if always factual—are too mundane to grab much attention and so anti-vaxxers rarely bother with them. Most of the attention in the media, spurred in part by anti-vaxxer claims, has been with respect to a much smaller set of side-effects.

We shall have to content ourselves here with four. With specific respect to COVID-19 vaccines it has been claimed that the following side-effects can be caused by the vaccine:

- immune system dysfunction;
- infertility;
- rewritten genetic code;
- stroke.

These are hardly minor side-effects and merit attention.

7.1 Immune System Dysfunction

In the previous chapter we saw claims made by anti-vax osteopath Sherri Tenpenny that COVID-19 vaccines actually cause the immune system to do more damage than good. Tenpenny references her own work, which makes claims about adverse effects of a COVID-19 vaccine by citing a study on SARS and MERS diseases.[204] The SARS-CoV virus and SARS-CoV-2 virus (which causes COVID-19) are both coronaviruses, but about a fifth of their genome is different. Tenpenny inappropriately, and without any supportive evidence, ties the study on SARS-CoV virus to the COVID-19 vaccine. She generalizes specific findings about limited effects on different viruses to the COVID-19 vaccines.

Now this isn't to say there is no grain of truth in what she reports. Scientists in Yale's Department of Immunobiology note that in *rare* cases the very pathogen-specific antibodies—like those against coronavirus' spike proteins—can actually promote pathology through a condition known as Antibody-Dependent Enhancement (ADE). This can, indeed, facilitate viral entry into certain cells (FcR expressing ones). But—and this is crucially ignored by Tenpenny—"There is no evidence that ADE facilitates the spread of SARS-CoV in infected hosts."[205] They further point out that multiple factors determine whether antibodies work in a protective or destructive fashion. For example, different kinds of vaccines tested on animals have produced different results. Thus, mice immunized using the inactivated whole SARS-CoV virus

---

[203] VAERS, "Disclaimer."
[204] Liu et al., "Anti-Spike IgC."
[205] Iwasaki and Yang, "The Potential Danger," 339.

could induce ACE. So this factor should be considered in COVID-19 vaccines.[206] But *none* of the approved American vaccines work in this manner (J&J uses an inactivated cold virus).

The above claim by Tenpenny is joined to others, such as that a COVID-19 vaccine can kill outright, or cause severe autoimmune diseases, or magnetize people, or somehow interface with 5G cell towers.[207] Facebook removed her page for spreading misinformation.[208] Tenpenny is another example of an anti-vaxxer with a profit motive. She manages an alternative health clinic and has made being an anti-vaxxer a business venture, producing anti-vax books, videos, and 'educational products.'[209]

7.2 Infertility

Pro-vaccine advocates find it very worrisome that anti-vaxxers have succeeded in sowing so much doubt among Americans. For example, an Ipsos/Axios poll in late March, 2021, found that more than one-third (35%) of Americans don't know whether the COVID-19 vaccine sterilizes recipients.[210] Clearly, then, this is a matter to which serious attention needs to be given.

With respect to *infertility*, we met again with Mike Yeadon, (with Wolfgang Wodarg) and a petition to Europe's medicines regulator to halt COVID-19 vaccine trials based on the assertion they could cause infertility in women. This suggestion was quickly picked up by many anti-vaxxers as a *fact* rather than as a *possibility*.

A team of researchers, using the tool Google Trends, investigated relative search volumes (RSV) for the terms 'infertility,' 'infertility AND vaccine,' and 'infertility AND COVID vaccine' for a one year period between February 4, 2020 and February 3, 2021. The petition was posted December 1, 2020 (i.e., more than nine months into the studied range). Peaks in searches for these terms then occurred during the two week period after the petition. In each case the actual RSV was significantly higher than the projected RSV. For example, in the case of 'infertility AND COVID vaccine' the actual RSV at peak was a 34,900% increase above expected RSV. The researchers observe such findings suggest the use by anti-vaxxers of this petition stimulated a surge in searches that reflected concerns over vaccine safety.[211]

The *fact* is that the protein targeted by the vaccine and syncytin-1 share a sequence of only four amino acids (of syncitin-1's 538 amino acids)—too few to prompt autoimmune response. But the alarm raised by the spectre of such harm was then joined to another line of supposed evidence for this argument—that of 23 women who became pregnant during Pfizer's vaccine trials, one experienced miscarriage, thus seemingly a 1-in-23 risk (4%). But the one woman

---

[206] Iwasaki and Yang, "The Potential Danger," 339–41.
[207] Blake, "Sherri Tenpenny's."
[208] Funke, "COVID-19 Vaccine."
[209] As stated on her Amazon author blurb.
[210] Ipsos/Axios Poll (March 19-22).
[211] Sajjadi et al., "United States Internet Searches."

who experienced this loss *received the placebo, not the vaccine*.[212] While the facts are inconvenient for anti-vax claims, they remain solid: no evidence supports a link between COVID-19 vaccination and infertility in either men or women.

7.3 Rewritten Genetic Code

A popular claim supported by anti-vaxxers like physicians Joseph Mercola (osteopath) and Andrew Kaufman (psychiatrist) is that a COVID-19 vaccine can rewrite the human genetic code.

Kaufman, who characterizes himself as "a natural healing consultant"[213] (i.e., *not* a member of the medical 'establishment')—thus linking himself to standard anti-vax cultural ideals—offers at his website a library of materials purporting "authentic medicine and freedom for all" (the latter being the central cultural ideal of the anti-vax movement).[214] We saw his claim that the COVID-19 'virus' is just an exosome, a particular genetic sequence randomly generated in the human body. But when the SARS-CoV-2 genetic code (less than 30,000 'letters' long) is compared to the human genome (more than 3 billion letters), it is found that the two share a miniscule 117 base pair stretch[215]—hardly the match Kaufman supposes. Moreover, the sequence in human beings looks like the virus' spike sequence—but it isn't viral. Remember, human cells possess antiviral proteins so when viral RNA is detected, it is shredded.[216]

But perhaps we should take more seriously the arguments of osteopath Carrie Madej. She warns the mRNA vaccines "are completely experimental," and use technologies "that can change the way we live, who we are, and what we are—and very quickly." Madej identifies the problem as 'modified RNA.' It isn't clear if she understands that the *m* in mRNA actually stands for 'messenger,'—not 'modified.' But at any rate her focus is on the delivery mechanisms for this 'modified' RNA: transfection, the enzyme Luciferase, and hydrogel.

Transfection refers to a process by which RNA is artificially introduced into an organism. But the word 'artificially' is hardly scary, since any man-made tool, from a hammer to a vaccine needle, is 'artificial.' Transfection provides a nonviral delivery mechanism through crafted mRNA and is a part of nucleic acid therapeutics. It has several advantages over older vaccine delivery models. For one thing, it is safe: "as mRNA is a non-infectious, non-integrating platform, there is no potential risk of infection or insertional mutagenesis." Second, it is more stable and highly translatable, meaning easily taken up and used in the host cells. Finally, mRNA vaccines are relatively inexpensive, can be rapidly produced, and done so at any desired scale of administration.[217]

---

[212] See both Reuters Staff, "Fact Check," and Fung, "7 Myths."
[213] Kaufman, "Bio & Credentials" (https://andrewkaufmanmd.com/bio-credentials/).
[214] To learn more about Kaufman, see Jarry, "The Physician Who Calmly Denies Reality."
[215] Lehrer and Rheinstein, "Human Gene Sequences in SARS-CoV-2," 1633.
[216] Lehrer and Rheinstein, "Human Gene Sequences in SARS-CoV-2," 1634.
[217] Pardi et al., "mRNA Vaccines." The quote is from p. 261. Also see Cao and Gao, "mRNA Vaccines," for a shorter discussion of the topic.

Luciferase enzyme, a natural bioluminescence found in fireflies, is *not* an ingredient of mRNA vaccines. (The list of ingredients is found elsewhere in this volume.) So that claim is absurd, though Madej—again without any justification—holds out the possibility this *might* be true in some *future* vaccine. As pointed out earlier, if one accepts that any *possibility*, even a *future* one, makes for a credible claim, then no claim is ever too incredible to be dismissed. That way of thinking, as a poet might say, leads only to madness.

Hydrogel, just as the name suggests, is a gel using water. They have been known since the late 19th century. Some are natural; some synthetic. They are widely used in biomedical applications because they provide for a controlled delivery of substances.[218] They have been suggested for mRNA vaccines because early studies indicate they inhibit cancerous processes.[219] But the American approved mRNA vaccines do *not* use hydrogel. They both employ lipid nanoparticles (i.e., very small ('nano') fat ('lipid') particles), which protect the mRNA from degrading before it can be used; these are easily taken up by cells, where they release their mRNA content.[220]

This brings up a related matter—the fear-mongering intended by speaking about 'nanotechnology.' Anti-vaxxers want people to associate the word with science fiction horror, such as tiny robots in the body running amok and taking over. But all the word means is 'very small technology.' It is used for many things and occurs commonly in medicine. The very small materials of nanotechnology—called *nanomaterials*, of course—can be natural as well as artificial. Nanotechnology has been very important in cancer treatment, diabetes management, and treating cardiovascular diseases—none of which seem to draw anti-vax ire. That nanotechnology would be used to combat viruses—very tiny particles themselves—is quite logical.[221]

As noted previously, the notion that a COVID-19 vaccine can alter a vaccinated person's DNA—rewrite the genetic code—has its genesis in the fact that the Pfizer and Modern vaccines are mRNA vaccines. We saw in an earlier chapter that RNA (ribonucleic acid) is not the same as DNA (deoxyribonucleic acid). The key element is *messenger* RNA (mRNA), which makes one specific protein and acts like a Snapchat message, providing a limited, specific message of very short duration. The message prompts a sequence in which the immune system learns to recognize the kind of viral profile COVID-19 presents and then responds by producing antibodies. To rewrite genetic code would require being able to penetrate a cell's nucleus, where DNA resides, which the mRNA is incapable of doing.

---

[218] Li and Mooney, "Designing Hydrogels for Controlled Drug Delivery."
[219] Liu, "An mRNA Vaccine Delivered in Hydrogel."
[220] "Let's Talk about Lipid Nanoparticles."
[221] For more on this big subject, see Arivarasan, Loganathan, and Janarthanan, *Nanotechnology in Medicine*.

7.4 Stroke/Blood Clots

As seen earlier, the VAERS database (as of early September, 2021) records 2,219 reports of a 'cerebrovascular accident' (.43% of all reported adverse events). The possibility of *blood clots* or *stroke* (specifically cerebral venous sinus thrombosis (CVST)) being caused by COVID-19 vaccines brought significant media attention in the early days of vaccine administration. This attention generated serious investigation. To repeat what we saw before, it was discovered is that after the first dose of a two dose vaccination there was some risk of CVST. Epidemiologist Julia Hippisley-Cox reports that of *10 million* people exposed to the AstraZeneca (ChAdOx1 nCoV-19) vaccine, there occurred *7* excess events of CVST in the initial 28 days after vaccination. With respect to AstraZeneca's vaccine—which has not been authorized in the United States—in Europe, of *34 million* recipients of the vaccine there were *53* possible cases of blood clots (specifically splanchnic vein thrombosis (SVT)), and at least *169* possible cases of CVST.[222] The raw numbers—53 and 169 or more—are frightening, especially out of context. But as a percentage of 34 million they represent a miniscule risk: 0.00016% for SVT and 0.0005% for CVST.

But what about the vaccines used in the U.S.? With respect to CNS blood clots (i.e., central nervous system thrombosis), of 54 million recipients of the Pfizer vaccine there were 35 *possible* cases (about 0.00006%); of the Moderna vaccine there were 5 possible cases out of 4 million recipients (0.000125%).[223]

What anti-vaxxers don't point out is that the risk of such events is *higher among the unvaccinated* who contract COVID-19. Hippisley-Cox observes that U.S. research found the risk of CVST associated with COVID-19 infection was about *seven times greater* than that associated with either the Pfizer of Moderna vaccines.[224] Or, to put it simply, as reporter Gabriela Miranda did for *USA Today*, "The risks of getting a blood clot if you contract COVID-19 is far greater than if you receive the Pfizer-BioNTech and AstraZeneca vaccines."[225] In fact, after reviewing this research Dr. Richard Francis, Head of Research at the Stroke Institute, remarked that the benefits of vaccination for *reducing* stroke risk from COVID-19 is so far greater than the risk of getting it from the vaccine as to justify vaccination on its own.[226]

Finally, we may also tag on to this argument one brought against vaccinations in general—that there are too many of them and the sheer number compound the danger. This argument appeals especially to those who know nothing of the history of vaccine development. It is true that more vaccinations are urged than ever before, but it is also true that advances in technology have meant the safe delivery of vaccines such that more than one often can be combined in a single treatment. This means that the total number of antigens—

---

[222] Hippisley-Cox, "Risk," 9–10.
[223] Hippisley-Cox, "Risk," 10.
[224] Hippisley-Cox, "Risk," 2.
[225] Miranda, "You're More Likely to Get a Blood Clot."
[226] Science Media Centre. "Expert Reaction to Study."

what people react to in a vaccination—is actually *smaller* than when vaccines were fewer in number.[227]

8. COVID-19 vaccines actually increase the risk of infection.

In the last chapter we saw how anti-vaxxers like to turn to VAERS to support the claim that vaccines increase the risk of infection as indicated by the numbers infected compared to the unvaccinated. As of September 8, 2021, COVID-19 itself was reported as an adverse event of vaccination 13,987 times (2.72% of all *reports*, not of all people vaccinated), and death was reported 5,118 times (1% of reports).[228] Once more the numbers need some meaningful context.

For example, Keziah Weir, Senior Editor at *Vanity Fair*, in an article on Robert F. Kennedy, Jr.'s claims, discusses the VAERS database. Weir points out that a large number of the death entries note the deceased was elderly and/or infirm, and many entries submitted by clinicians presume the recorded death was unrelated to the vaccine. Weir points out that many factors are observed in cases, such as the person committing suicide or already having been infected when vaccination was begun. In one case the deceased was 99 years old and had refused all food for a week and died within 12 hours of receiving the vaccine.[229] This kind of context doesn't serve the anti-vax agenda and so is left aside.

As we saw, the idea that vaccines actually *increase* the risk of getting COVID-19 sometimes plays out in anti-vax claims in two ways: that one can contract COVID-19 from the vaccine itself (or a disease-like effect that is the same), or that vaccinated people get sick more often than unvaccinated ones.

8.1 Contracting COVID-19 after Vaccination and/or from It

This first line of thinking sometimes seems to confuse vaccination with variation. The latter is the use of a small amount of the disease to give a person a mild case of the disease and thus inoculate them against more severe infection. This was the sort of thing once done with smallpox before Edward Jenner introduced the use of cowpox in place of smallpox. *COVID-19 vaccines have no live COVID-19 virus so they cannot cause the disease.*

But people can contract COVID-19 after vaccination—so-called 'breakthrough' infections. A vaccine does *not* guarantee immunity; it merely decreases the probability both of infection, and of serious infection, so dramatically as to seem like complete immunity to some people. This matter is addressed enough elsewhere in this volume as to need no more comment here.

However, we saw the claim made really doesn't have breakthrough infections in mind, but instead argues the vaccine itself is the cause of ill-health. In this respect, the vaccine doesn't have to cause COVID-19 if its effect is the

---

[227] Offit et al., "Addressing Parents' Concerns." They document that children receive *fewer* antigens from vaccinations today than was true a century ago—or even forty years ago.
[228] VAERS (https://wonder.cdc.gov/controller/datarequest/D8; jsessionid=26BE0094FD4089DEDA988EC90098), Sept. 8, 2021.
[229] Weir, "How Robert F. Kennedy Jr. Became the Anti-Vaxxer Icon."

same as the disease. This is the argument of the *Vaxxter* article that claims as its support a paper in the journal *Circulation Research*. The anti-vax claim is that the Spike protein itself is introduced by mRNA vaccines (Pfizer's and Moderna's), and acts just like the SARS-CoV-2 protein, doing significant damage. But this misrepresents the facts, both with respect to what mRNA vaccines do, and with respect to what the article actually says.

First, mRNA vaccines do *not* inject viral Spike proteins. What the vaccine does is teach the human cell how to make such a protein so as to recognize it if introduced by a virus. Seeing what a cell has made prompts the immune system to recognize it as foreign and unwanted, then develop antibodies against it. But both Pfizer's and Moderna's vaccines also contain two stabilizing mutations designed to prevent certain undesired changes.[230] So the fears generated by the *Vaxxter* article are groundless. To use a metaphor expounded earlier in this book, the vaccines provide basic training to new troops to prepare them if war erupts, but they don't use live ammunition.

Second, the journal report appealed to by *Vaxxter* does not conclude what the anti-vax article wants it to. The study reported research done with Syrian hamsters, which found that the S protein can downregulate ACE2 receptors, inhibiting mitochondrial function, and thus damage vascular endothelial cells, even though they also decrease viral infectivity. The authors recognize the need in vaccination to guard against this—which we just saw mRNA vaccines do—and thus they explicitly say, "vaccination-generated antibody and/or exogenous antibody against S protein not only protects the host from SARS-CoV-2 infectivity but also inhibits S protein-imposed endothelial injury."[231]

8.2 Vaccinated People Get Infected More than Unvaccinated Ones

The second claim—that vaccinated people get infected more than unvaccinated ones—illustrates an old saying in teaching statistics, one I used with every class of students when I taught such courses: "Figures don't lie, but liars figure." Raw numbers don't usually tell us much; they require context. That is the case here.

Historically vaccines have been so successful that near universal vaccination almost has eliminated once common diseases like measles. But when outbreaks occur, as they did recently in the U.S., both vaccinated and unvaccinated people are at risk. In fact, in terms of raw numbers, *more* vaccinated people are likely to contract a disease than the unvaccinated in a heavily vaccinated society. Therefore, runs this argument, don't get vaccinated because it increases risk.

But this ignores the context. In a case like measles, most Americans have been vaccinated. Let us suppose a town of 1,000 residents, in which 990 have been vaccinated with a success rate of 97% (the actual rate of success for

---

[230] Heinz and Stiasny, "Distinguishing Features of Current COVID-19 Vaccines."
[231] Lei et al., "SARS-CoV-2 Spike Protein." Also see Yang and Du, "SARS-CoV-2 Spike Protein," where they conclude (p. 3), "[T]he S/RBD-specific antibodies have great potential fordevelopment as effective therapeutics that prevent COVID-19 spread." The issue is not so much the targeting of the S protein, but *how* it the human host cells are taught to respond to it.

measles). Now an outbreak occurs. The number infected turns out to be 39 people—and 30 (roughly 3% of the 990) of them are vaccinated! That is more than four times the number of those who were never vaccinated. Presented as raw numbers (30 to 9), with an accompanying statistic (77%), it looks like a good case for staying away from the vaccine. But it conveniently ignores that 90% of the unvaccinated were infected (the actual rate for unvaccinated with exposure). Those vaccinated had very little chance of infection, while the unvaccinated had a very high chance (90% for the latter as opposed to 3% for the former).[232] No one needs be able to do more than basic math to figure out what the real odds are of getting sick.

This ignorance of context, or willful avoidance of considering it, has plagued anti-vax credibility. It shows up often in anti-vax presentation of 'data.' For example, let's reconsider McBean's *The Poisoned Needle*. She presents impressive numbers and quotes from authorities—*but absent context*. McBean bolsters numbers with selective, out-of-context quotations. For instance, her first supportive quote is from the recorder for the city of London who reported that of 155 persons admitted to a hospital with smallpox, 145 of them had been vaccinated.[233] Again, this looks compelling: 93.5% of hospital admissions were vaccinated people! But look back at our example above. We should expect in a heavily vaccinated populace that most cases would be vaccinated folk. The really relevant data is what percentage the 145 and 10 represented of their respective groups, the vaccinated and unvaccinated. But this crucial contextual data is left aside.

This kind of fiddling with data and sources is why subsequent researchers have not endorsed her work. In fact, Professor of Medicine Gareth Williams, in his history of smallpox, notes that McBean also was fond of saying 'Figures cannot lie, but liars figure,' but he adds, "True, but it was the anti-vaccinationists who had the worse record for torturing statistics into making false confessions."[234]

We may footnote this matter of the history of smallpox vaccine—so critical a source for anti-vaxxers—by recalling attention to the actual success of the vaccine as documented across time by the mortality rate. In England, before vaccination, 3,000 per million people died of smallpox. From 1838–1853, when vaccination was optional, the rate declined to 417 per million people. From 1857–1866, when vaccination became mandatory, the rate fell further, to 214 per million. From 1889–1898, when vaccination was vigorously enforced, the

---

[232] See the discussion at *The History of Vaccines* website (https://www.historyofvaccines.org/content/articles/misconceptions-about-vaccines), sponsored by the College of Physicians of Philadelphia.
[233] McBean, *The Poisoned Needle*, 11.
[234] Williams, *Angel of Death*, 285.

rate plummeted to 10 per million.[235] It is a remarkable stretch to call the smallpox vaccine a failure, let alone a threat to health.

In the last chapter we saw the case of the outbreak in Barnstable County (Provincetown), Massachusetts, where nearly three-quarters (74%) of those contracting COVID-19 in an outbreak were vaccinated. But like the smallpox history, this statistic needs context. The vaccination rate in Massachusetts at the time was 69%. The outbreak was linked to a so-called 'superspreader event'—thousands of tourists converging on a town. The impact is reflected in the before and after number of cases of COVID-19. *Before* the event (July 3), 0 cases per 100,000 persons were recorded in the county; *after* the event (July 17, the next reporting time), 177 cases per 100,000 people were recorded. Among the 346 breakthrough cases, only 4 were serious enough to require hospitalization (1.2%), and none of them died.[236] Boston epidemiologist Matthew Fox has remarked, "Provincetown is an area with some of the highest vaccination rates in the country, so if the vaccine was not working, you'd expect the % vaccinated among the infected to be even higher than 75%. . . ."[237] As we saw above, it is expected that where a majority of people are vaccinated the incidence of infection will be higher than among the unvaccinated. Perhaps better takeaways from this incident are, first, that COVID-19 (in this case the delta variant) is highly infectious, and second, that the vaccinated had enough protection to avoid serious illness and death.

Even more important to note, though, is that the anti-vax claim that the vaccinated are as likely to spread COVID-19 as the unvaccinated ignores the most salient contextual fact: *to spread COVID-19 one must have it, and the unvaccinated are much more likely to have it than the vaccinated*. So, like some other anti-vax claims, this one has a small measure of truth. Those infected with the disease can spread it. But who is more likely to be infected in the first place?

9. Other, better courses of action exist.

In the previous chapter we saw claims with respect to a half-dozen alternatives.

9.1 Prevention Strategy: Vitamins C and D, and Zinc

We shall start by revisiting what was said in the last chapter. Vitamins and supplements like zinc are often recommended by health professionals—one area where anti-vaxxers and medical professionals actually might agree. *Vitamin C* (ascorbic acid) is perhaps the most popular vitamin people take as a supplement to their diet. It is considered an essential nutrient and is important to the body in many ways, including boosting the immune system. *Vitamin D* is probably most famous for its role in bone health, but is also thought to benefit the immune system by increasing antimicrobial peptides the body uses as a

---

[235] Statista, "Average Number of Annual Smallpox Deaths." Statista is a company specializing in providing statistical data.
[236] Brown et al. "Outbreak of SARS-CoV-2 Infections."
[237] Fox is quoted in McDonald, "Posts Misinterpret."

natural antibiotic and to be valuable in protecting against respiratory infections. *Zinc*, too, supports the immune system, as well as boosting metabolic function. Thus all three are beneficial when taken in appropriate amounts.

Of the three, vitamin D has the best support for possessing value against COVID-19 infection. Research at the University of Chicago Medicine found that having vitamin D levels above those usually regarded as sufficient (30 ng/ml) may lower the risk of COVID-19 infection, especially for African-Americans. Levels of 40 ng/ml or greater seemed to decrease risk of infection. Moreover, vitamin D deficiencies (<20 ng/ml) seem associated with greater vulnerability to COVID-19. Yet these results were not found for Whites.[238] Since African-Americans as a group are more reticent about receiving medical care, this information might be especially useful for them.

Vitamin D as a dietary supplement is recommended at 600–800 IUs a day, with most people tolerating up to 4,000 IUs a day, but over 10,000 is probably unsafe. And before anyone rushes out to stockpile it, notice should be taken that a genetic study undertaken to eliminate the kind of confounding variables that make studies like the one done at the University of Chicago Medicine hard to test for verifying results found no evidence to suggest that taking vitamin D supplements would help against COVID-19.[239]

There is no evidence to date suggesting that vitamins C or D, or zinc, can either prevent or treat COVID-19. Moreover, it is possible to overdose with something like zinc, inducing headache, nausea and vomiting, and other unpleasant consequences—and all without achieving either any protection from COVID-19 or curing it. That said, pro-vaccine advocates have no argument against the appropriate use of supplements; they only object to misleading claims about them or the abuse of them.

9. 2 Treatment Strategy: Early Treatment

Anti-vaxxers, or at least those sympathetic to their cause like the Association of American Physicians and Surgeons (AAPS), argue that early treatment strategies are being ignored or at least downplayed by pro-vaccine advocates because of their preoccupation with prevention. This is perhaps the most credible argument an anti-vaxxer can make, though it is not really one against vaccines *per se*. Pro-vaccine advocates also support early treatment. Where they disagree with anti-vaxxers is what constitutes appropriate early treatment. Anti-vaxxers want the right to choose powerful drugs that have not been demonstrated as safe and effective for treating COVID-19. Pro-vaccine advocates resist such choices not as a matter of individual conscience but as a matter of public policy. Thus the difference here between the two groups is principally a reflection of their differing cultural values.

---

[238] Caldwell, "Study."
[239] Butler-Laporte, et al., "Vitamin D and COVID-19."

9.3 Prevention/Treatment Strategy: Hydroxychloroquine

Hydroxychloroquine (HCQ) presupposes infection with COVID-19; it is *not* preventative in nature, but a treatment for after infection. As we saw earlier, because it has anti-inflammatory properties it can throttle down an over-active immune response. Such factors suggested that hydroxychloroquine might be an effective way to treat COVID-19.[240] In fact, on March 28, 2020, the FDA authorized emergency use of hydroxychloroquine to treat COVID-19. But that authorization was subsequently revoked a short time later (June 15, 2020).

The anti-vax doctor Jane Orient, in testimony before a Senate committee made six basic arguments (summarized in the last chapter). Only one directly had to do with HCQ; she argued briefly that it is safe and contended it should be "legitimized" for use "outside of a hospital setting based on a physician's judgment." Those words highlight the cardinal theme of the Association of American Physicians and Surgeons (AAPS) that she represents. The remainder of her arguments reiterate AAPS cultural (political) convictions: that the government agency FDA overreaches its authority and restricts the practice of individual physicians; that its evidence-based standards for approving treatments is excessive and interferes with individual doctor's judgments; and that appeals to studies in prestige journals is no guarantee of 'reliability' because such journals get fooled, too (and she alludes to *The Lancet* scandal of publishing, then retracting the Wakefield et al. paper linking vaccination to autism). As is typical of anti-vax arguments, science takes a back seat to cultural concerns.

FDA authorization was revoked because further research found no benefit with respect to either speeding patient recovery or lowering the risk of death. Two weeks later (July 1, 2020) the FDA issued an update specifying the risks of serious heart rhythm problems, blood and lymph system disorders, kidney injuries, and liver problems, including liver failure.[241] In short, this proved to be a case where the treatment could very well be more dangerous than the disease.

9.4 Treatment Strategy: Remdesivir

Hope has also been held out for *remdesivir*, an anti-viral medication administered by injection. It operates by blocking a virus from replicating itself, specifically by interfering with a key enzyme the virus requires for replication. In a double-blind, randomized, placebo-controlled study involving 1,062 hospitalized COVID-19 patients, remdesivir was a superior treatment compared to placebo. A 10 day course of the drug had shorter recovery times and might also have prevented progression to more severe respiratory disease.[242]

This led to emergency authorization by the FDA (May 1, 2020) to use remdesivir to treat hospitalized COVID-19 patients. Subsequently, the FDA expanded its authorization in October, 2020, by "no longer limiting its use to

---

[240] Sinha and Balayla, "Hydroxychloroquine and COVID-19."
[241] FDA, "FDA Cautions against Use of Hydroxychloroquine."
[242] Beigel et al., "Remdesivir for the Treatment of COVID-19."

the treatment of patients with severe disease." This expanded authorization also allows health care facilities other than hospitals to offer the treatment, provided they are "capable of providing acute care comparable to inpatient hospital care."[243]

The Infectious Diseases Society of America (IDSA) notes that the largest randomized trials of remdesivir showed varying results. One, the Adaptive COVID-19 Treatment Trial (ACTT-1), involving 1,062 patients, found that remdesivir administered at a median time of 6 days after symptom onset shortened recovery time from a median time of 15 days to a median time of 10 days. Compared to a placebo group the mortality rate was 11.9% to 6.7% for the remdesivir-treated group.[244] On the other hand, the SOLIDARITY trial, a much larger study, with more than 11,000 patients, found a mortality rate of 11.8%, with no difference between groups receiving remdesivir or not.[245] So, in 2021, a WHO guideline committee recommended against its use. The reasoning—which extends to hydroxychloroquine, too—is that neither drug significantly reduces the likelihood of death for those hospitalized with COVID-19.[246]

9.5 Prevention/Treatment Strategy: Ivermectin

More recently, *ivermectin* has been championed as a vaccine alternative. Conservative politicians like Republican Senator Ron Johnson (Wisconsin) and Fox News Host Laura Ingraham publicized and endorsed its use for COVID-19. This provided 'authoritative' cover for anti-vaxxers, especially when Ingraham touted the conspiracy theory that drugs like ivermectin are being "suppressed" to protect the approved vaccines.[247] Plus, a preprint scientific paper claimed to demonstrate that ivermectin can reduce COVID-19 deaths by 90% or more.[248]

Unlike remdesivir and hydroxychloroquine, ivermectin is being touted by anti-vaxxers as a *preventative* medication and, if needed, a treatment after infection. Antibacterial in nature, ivermectin is commonly employed as a preventative medication against heartworm is some small animal species. It is also FDA approved for human use to treat certain parasitic worms and as a topical treatment for external parasites like head lice.[249]

As an *antiparasitic* drug, ivermectin has not been approved for *antiviral* use. However, some studies suggest it might be effective against viruses because in laboratory settings it has been seen to interfere with the process viruses use to enhance infection and inhibit an immune response. With specific respect to the COVID-19 virus, ivermectin might interfere with the spike protein it uses to attach to human cells. The problem, though, is that studies also suggest that an

---

[243] Hinton, "Letter to Ashley Rhoades."
[244] Beigel et al., "Remdesivir for the Treatment of COVID-19."
[245] WHO Solidarity Trial Consortium, "Repurposed Antiviral Drugs for COVID-19."
[246] Miller, "Remdesivir, Hydroxychloroquine."
[247] Blake, "How the Right's Ivermectin Conspiracy."
[248] Elgazzar, et al. "Efficacy and Safety of Ivermectin."
[249] FDA, "FAQ: COVID-19 and Ivermectin."

effective dose of ivermectin to treat COVID-19 would need to be up to *100 times higher* than what is currently approved as safe for human beings. Those studies that have been done with people who have COVID-19 have met with mixed results.[250] In fact, the very research that anti-vaxxers thought offered strong support for ivermectin when released as a preprint, never saw publication because it was withdrawn after serious questions concerning plagiarism and data manipulation indicative of ethical violations of scientific standards.[251]

However, the study cited in the last chapter concerning administration of oral ivermectin producing a protective anti-viral effect is an encouraging one. But the authors of that report noted ivermectin's effectiveness as an anti-viral operates within a therapeutic margin (meaning dosage for effectiveness is a critical factor).[252] Thus they only recommend further investigation, not a wholesale rush to ingest large quantities.

In the Spring of 2021, the World Health Organization (WHO) stated, "The current evidence on the use of ivermectin to treat COVID-19 patients is inconclusive." It thereby recommended that it be used only in clinical trials to further study its possible benefits and/or risks. This recommendation was made after reviewing pooled data from 16 randomized controlled trials involving more than 2400 patients in both inpatient and outpatient settings.[253]

In the United States, at the beginning of September, 2021, the American Medical Association (AMA), American Pharmacists Association (APhA), and the American Society of Health-System Pharmacists (ASHP) issued a joint statement in which they declared that they "strongly oppose the ordering, prescribing, or dispensing of ivermectin to prevent or treat COVID-19 outside of a clinical trial." They note that even the drug's manufacturer (Merck) says there is insufficient evidence to support its use to treat COVID-19.[254]

Ignorant of the full range of needed facts, fed by fears of vaccination and spurred on by anti-vax enthusiasm, a number of people have turned to over-the-counter veterinary doses of ivermectin intended for horses and cows. Unfortunately—if predictably—the result has been a dramatic increase in overdoses. The National Poison Data System (NPDS) reported a 245% increase from July to August of 2021, with more than a thousand cases from the start of the year to the end of August, itself a 163% increase from the same period in 2020. Overdosing can produce symptoms such as dizziness and balance problems (ataxia), nausea and vomiting, diarrhea, lowered blood pressure

---

[250] NIH, "Ivermectin."
[251] For more, see Reardon, "Flawed Ivermectin Preprint." The withdrawn paper was Elgazzar, et al., "Efficacy and Safety of Ivermectin."
[252] Bernigaud, "Oral Ivermectin," 1208.
[253] WHO, "WHO Advises."
[254] AMA, "AMA, APhA, ASHP Statement." So much for the anti-vax argument that all the pharmaceutical companies care about is profit. If that were so Merck should be at the front of the line urging the use of ivermectin.

(hypotension), and allergic reactions. Worse, it can lead to seizures, coma, and death.[255]

Let the FDA have the last word on this one: "You are not a horse. You are not a cow. Seriously, y'all. Stop it."[256]

A Quick Final Warning

In September, 2021, it was reported that some people were turning to *gargling iodine* as a preventive measure against COVID-19.[257] The manufacturer, Betadine, published an online statement on the matter: "Betadine® Antiseptic Sore Throat Gargle is only for the temporary relief of occasional sore throat. Betadine Antiseptic products have not been demonstrated to be effective for the treatment or prevention of COVID-19 or any other viruses."[258] The adaptation of a drug for some other, nonprescribed use, is an inherently risky move—more risky than any vaccine.

10. *COVID-19 will decline without the need for vaccines, etc.*

The last argument we will consider is actually two-fold: that coping with infectious diseases need not require any human intervention because they decline naturally, and that natural immunity is better anyway. Let's look at each part of this argument.

10.1 Diseases Decline on Their Own.

The first contention is that diseases decline on their own without the need for vaccines.

Unfortunately, the idea that a serious viral disease like COVID-19 will go away on its own is true only in a heartbreaking way. After it spreads widely enough to kill many more people it might eventually mutate into a variant that *only* makes people sick without also killing many of them. Then it will seem to have 'gone away.' By then it really will be like the 'flu' so many anti-vaxxers have compared it to. A similar thing happened with the 1918 influenza, which was one of the great killers of history. It eventually mutated into less deadly strains. On the other hand, it might instead prove to be like the smallpox virus that persisted a virulent killer for centuries until a vaccine stopped it, and eventually eradicated it. Betting on natural decline is cold-comfort to the dead and those who mourn them.

With respect to the anti-vaccine arguments in general on this point it must be said that they typically rely on misinterpretations of epidemiological data or selective use of statistics, such as reporting mortality rates (i.e., how many die) without also looking at incidence rates (i.e., how many new cases of infection

---

[255] Romo, "Poison Control Centers."
[256] Twitter tweet from FDA (https://twitter.com/us_fda/status/ 1429050070243192839)
[257] Dawson, "Anti-vaxxers Are Gargling Iodine."
[258] Betadine, "Betadine Antiseptic Sore Throat Gargle Usage Guidelines" (https://betadine.com/covid-19/#1596710758978-f453fb55-1908)

occur in a specified period of time).[259] This sort of fiddling with the data is what we just reviewed in considering the previous anti-fax argument.

No one size fits all with disease histories. Looking to past instances of diseases is of only limited assistance with COVID-19. Although coronaviruses have been known and studied for almost a decade, no serious diseases from them emerged until the 21st century. Since we are less than a quarter of the way into this century there is no way to know or even to accurately predict whether SARS, MERS, or COVID-19 will prove scourges lasting centuries or be old news a decade from now. The anti-vax argument of an inevitable decline is simply wishful thinking.

10.2 Natural Immunity Is Superior to Vaccine Immunity

The second contention is that natural immunity (i.e., immunity through contracting the disease) is superior to the 'artificial' immunity of vaccination. Now let's pause before we look at how this argument is made by considering it in light of what we have already seen, including just above. Diseases like smallpox, left to spread through populations to achieve 'natural immunity,' persisted for centuries as widespread and deadly entities. Vaccination progressively reduced smallpox's power by achieving a degree of herd immunity that natural means never achieved. Any number of infectious diseases show a similar history. If natural immunity is the better path to herd immunity, why doesn't history show so?

One place mentioned in the last chapter is the anti-vax website DangersofVaccines.com, operated by health coach Landee Martin and network marketing professional Walt Shilinsky. Unfortunately, little of the material actually addresses the claim that natural immunity is superior. The only place where the central claim is actually addressed is when the website declares that getting sick confers lifelong immunity whereas vaccines do not.[260] None of these claims is substantiated by supporting evidence beyond vague references to a short news article about the decline in measles infections, an article claiming a woman who died of measles had been vaccinated, and the opinions of a retired geologist (Viera Scheibner) voiced in an anti-vax movie. The pediatrician quoted at the head of the material had his license to practice medicine suspended for fraudulent assertions constituting an immediate danger to the public.[261]

We next considered another source. Daniel Erickson of America's Frontline Doctors, with his partner Artin Massih, who argued that when people shelter away from COVID-19 it *decreases* the immune system with the result that when exposure does occur diseases spike.[262] This is simply a nonsensical claim. Of course exposure will increase disease incidence, but that doesn't follow from a decreased immune response; it follows from increased exposure. Avoiding

---

[259] See Camargo Jr., "Here We Go Again," 4.
[260] "Natural Immunity vs. Artificial Immunity."
[261] On Paul Thomas, MD, see "Pediatrician's License."
[262] See their comments at Brownstein, "COVID-19 Lockdowns."

serious infections is hardly a bad idea. These doctors are not motivated so much by a science of immunology, but by their cultural and economic interests.

(While we are thinking about Erickson's claim we might take a moment to consider the group he affiliates with. America's Frontline Doctors, founded by physician Simon Gold, a political activist arrested for his participation in the riots at the U.S. Capitol on January 6, 2021, views the matter of the pandemic predominantly through a conservative cultural lens. They argue that COVID-19 is not a health emergency, that the national response to it is overblown, that the vaccines are ineffective, and that in the event of infection hydroxychloroquine is recommended. They had a website, but the webhost (Squarespace) shut it down, citing concerns over 'activity that's false, fraudulent, inaccurate, or deceiving'—and Facebook, Twitter, and YouTube all deleted its video-recorded press conference on COVID-19.[263])

Perhaps the most sophisticated anti-vax argument for natural immunity we saw was this: viral infections like COVID-19 provoke a natural immune system response; the more people who are infected, the better, because then herd immunity results and the virus is stopped. Most people either do not get sick (asymptomatic), or are not very seriously sick. Infected people who are not very sick spread the disease rapidly among other low-risk people and herd immunity then blocks the virus from infecting those most vulnerable to it.[264]

We saw that this is the line of reasoning espoused by physician Scott Atlas. His arguments may be motivated by cultural concerns related to politics and business, but setting aside such matters let us consider the logic. In essence he argues that *because* most people won't die, let them get infected, *because* that will promote natural herd immunity and *therefore* end the crisis. We need not contest this logic to point out that it is cold comfort to the infected who do die or end up with long-term effects, or who lose loved ones to the virus. It is, as some have put it, playing Russian roulette with COVID-19 and betting your number won't come up. But it has for millions of people around the world.

Finally, we saw the anti-vax website imposingly named "National Vaccine Information Center" (the creation of Barbara Loe Fisher and Joseph Mercola[265]) claim "Research on natural immunity from SARS-CoV-2 infection varies and suggests that durable immunity to the virus lasts for at least eight months and may be life-long." Three sources are cited in support.

The first is titled "Immunological Memory to SARS-CoV-2 Assessed for Up to 8 Months after Infection." The title alone refutes the claim made that durable immunity lasts for at least eight months; the study was only assessing that period of time. The research is a study of B cells, CD8+T cells, and

---

[263] See Price, "Controversial Bakersfield Doctor," and Melendez, "Demon Sperm Doc's Pals." (The title is referring to pediatrician and minister Stella Williams, who has claimed gynecological problems are caused by having sex with demons and witches in dreams.)
[264] See Atlas, "The Data Is In."
[265] Mercola has contributed more than $2.9 million dollars. See Jarry, "The Anti-Vaccine Movement in 2020."

CD4+T cells, all of which play a role in so-called 'immune memory,' and all of which operate in relative independence of one another. The study examined 188 people (108 female; 80 male) presenting with a range of asymptomatic, mild, moderate, and severe COVID-19 cases. The researchers explicitly remind us, "It is well recognized that the magnitude of the antibody response against SARS-CoV-2 is highly heterogeneous between individuals. We observed that heterogeneous initial antibody responses did not collapse into a homogeneous circulating antibody memory; rather, heterogeneity is also a central feature of immune memory to this virus." In other words, immune response to SARS-CoV-2 is *highly variable among people*. No universal claim can be made about how all people will have the same or similar antibody response, *or* immune memory. Further, the authors observe that both the severity of COVID-19 and factors like gender contribute to heterogeneity in immune memory. The bottom line, they say, is that "the source of much of the heterogeneity in immune memory to SARS-CoV-2 is unknown and worth further examination."[266]

The second study cited, in support of the claim 'and may be life long,' is titled "SARS-CoV-2 Infection Induces Long-lived Bone Marrow Plasma Cells in Humans." Once more the title should at least tip one off that the article is making a rather limited and specific claim, not a global one. The authors early on, with the existing scientific literature in mind, state, "Reinfections by seasonal coronaviruses occur 6 to 12 months after the previous infection, indicating that protective immunity against these viruses may be short-lived. Early reports documenting rapidly declining antibody titres in the first few months after infection in individuals who had recovered from COVID-19 suggested that protective immunity against SARS-CoV-2 might be similarly transient." Their own study was both limited in scope—19 convalescent individuals who had experienced mild infection—and focused in targeting bone marrow plasma cells. Their interest in such cells stems from the role they play in supplying antibodies. They studied bone marrow samples ('aspirates') from the 19 participants and in 15 of them detected SARS-CoV-2 S-specific BMPCs (Bone Marrow Plasma Cells), indicating humoral immune memory at 7 months after infection. This is sufficient to be termed "long lived"—the term seized upon by anti-vaxxers—and indicates that alongside S-binding circulating memory cells the human immune system has a robust response to SARS-CoV-2.[267] This is good news, but it is *not* the same as saying one thus acquires life-long immunity after a mild infection.

Finally, the third study cited, also in support of the claim of life-long immunity from infection, is titled "Comparing SARS-CoV-2 Natural Immunity to Vaccine-induced Immunity: Reinfections Versus Breakthrough Infections." This is a preprint article (i.e., before peer review) and compared three groups in Israel: one had received both doses of the Pfizer vaccine and had been

---

[266] Dan et al., "Immunological Memory." The quotes are from pp. 8, 9, respectively.
[267] Turner et al., "SARS-CoV-2 Infection." The quote is from p. 421.

uninfected (the reference group), another was unvaccinated and had been previously infected with SARS-CoV-2, and the third had both been previously infected and received a dose of vaccine. The study acknowledges that short-term effectiveness of the vaccine has been demonstrated; they were interested in its longer term effectiveness when compared to natural immunity via infection. They note their study was done at a time when the delta variant predominated and that long-term effectiveness across variants is unknown. They note the long-term effectiveness of previous infection remains unknown, too—an important point anti-vaxxers like to ignore. Their research found a significant advantage in long-term protection via natural infection. They speculated this might be because the immune response is to actual SARS-CoV-2 spike proteins rather than to the artificial anti-spike protein immune activation spurred by the vaccine. But they emphasize this is just a hypothesis. After noting several limitations to their study they conclude that natural immunization does seem to confer "longer lasting and stronger protection" than vaccination, *but* they also report that vaccination after infection provides even better protection.[268] Thus they are *not* supporting vaccine resistance. This study does provide support for an argument that natural immunization may be stronger than vaccine immunization, but that is not the same thing as saying it is automatically life-long and that vaccination is not advantageous. The study says neither of those latter things.

So let's review some of 'immunology 101' that anti-vaxxers Erickson and Massih refer to. We can start with the idea that the immune system is built by exposure to environmental agents. That is true. Our innate immune system—the one we are born with—becomes paired with an *adaptive* immune system through such exposure. Thus the human immune system is dual: both an innate system and an acquired one. The key to understanding how it works, though, is recognizing that it responds to *any* environmental agent it perceives as a threat, whether that agent is from nature or technology. Thus a response can come from either COVID-19 itself or from a COVID-19 vaccine.

Let's review a key point made in the pro-vaccine chapter. A tremendous advantage is provided by vaccination over relying on infection to achieve 'immunization.' In both cases a human body begins with *no* antibodies to the SARS-CoV-2 virus that causes COVID-19. The vaccinated body, though, pre-emptively begins making antibodies so that if and when the virus is encountered, *the body is already prepared to fight.* The unvaccinated body must react on-the-fly, with no previous preparation. It is disadvantaged at the start.

It is undoubtedly correct that sometimes misguided human zeal leads to a misuse of technology such that people try to shield themselves, for example, from *all* bacteria rather than distinguishing healthy bacteria from unhealthy ones. But it is a substantial leap from acknowledging that to arguing, 'Therefore, let us expose ourselves willingly to whatever comes along.'

---

[268] Gazit et al., "Comparing SARS-CoV-2 Natural Immunity to Vaccine-induced Immunity."

With the COVID-19 virus it seems there are three alternatives:
- ❖ Avoid exposure;
- ❖ Accept exposure and infection in the name of achieving 'natural immunity'; or
- ❖ Accept vaccination as a different way to achieving immunity.

It seems fair to guess most people like the first alternative best. But it has proven difficult to achieve. So, most people reckon that one or the other of the acceptances—of infection or vaccination—is probably inevitable. So which is preferable? We've seen the anti-vax argument and we know from the grim statistics the price being paid for accepting the path of 'natural immunity' through infection. It is like playing Russian roulette; yes, there may be more empty chambers (i.e., a slim chance of serious illness or dying), but that is no comfort to the unlucky ones. And as the sheer volume of infections, serious illnesses, and deaths show, a lot of people are unlucky.

Let's pause a moment to deal with the term 'immunity.' It is a myth that either vaccination or infection with COVID-19 guarantees immunity in the sense of 'No more worries folks! I'm completely protected forever!' No matter which of these alternatives one chooses, subsequent infection by the virus is possible. All the available evidence indicates that 'immunity,' whether conferred by infection or vaccination, wanes over time.

The fact is, even if we hypothesize that herd immunity to the COVID-19 virus is achieved with 50% population immunity, we don't know how long such immunity lasts. Researchers Arnaud Fontanet and Simon Cauchemez point out that immunity to season coronaviruses is usually short-lived, and this may be even more the case for those who contract mild cases of infection. Thus it might take repeated cases of re-infection to achieve the kind of strong protection everyone desires.[269] Moreover, as one recent study warns, the rapid viral adaptations shown by the COVID-19 virus (SARS-CoV-2) mean it is possible herd immunity will *never* be achieved.[270]

Let us, for the sake of argument, grant that the protection afforded by actual infection is superior to that conferred by vaccination. It may well be; various studies indicate an advantage. But that presupposes *surviving* the infection in the first place, and then also *avoiding the long-term effects and deficits* that often follow infection, and having been *sick enough the first time* to acquire strong and lasting immunity. All three conditions are important to consider if choosing the path of infection to achieve natural immunity. A study on herd immunity through a policy of allowing a large fraction of the populace to become infected (as advocated by anti-vaxxers) warns, "Unchecked, the spread of SARS-CoV-2 will rapidly overwhelm healthcare systems. A depletion in healthcare resources

---

[269] Fontanet and Caucheme, "COVID-19 Herd Immunity."
[270] Dopico, et al., "Immunity," 12.

will lead not only to elevated COVID-19 mortality but also to increased all-cause mortality."[271] And isn't that exactly what we have seen in the United States?

A significant complicating factor is that so-called natural immunity tends to be much more variable than that found with vaccination. In other words, as mentioned earlier, it is much more difficult to predict an individual immune response compared to the far greater homogeneity produced by vaccination.[272] Put in slightly different terms, those who are vaccinated tend to look much more alike in their immune system profile (a relatively 'homogenous' response) than those who are infected (a 'heterogeneous' response). That means *one individual's immune response to a natural infection may confer better subsequent protection compared to what he or she might have achieved by vaccination, but another individual's response might be far poorer.*

There may be some advantages to the protection gained by infection rather than vaccination, but it is best understood in terms of *large groups* of people rather than individuals. A study reported in preprint examined natural infection versus vaccination by the mRNA vaccines (Pfizer's and Moderna's). It found that vaccination produces greater amounts of circulating antibodies than natural infection. On the other hand, actual infection may result in a greater advantage down the road when encountering virus variants. The study's authors suggest those vaccinated could profit, then, from subsequent booster shots.[273] That is the same thought pointed out by the authors of the Israeli study.

In fact, the idea that those who have been infected can profit from subsequent vaccination and should be vaccinated has been shown in research done in this country. A study conducted in Kentucky found that unvaccinated survivors of COVID-19 are twice as likely to contract it again as people fully vaccinated after initially contracting the virus.[274]

What are we left with? Natural immunity appears to provide superior protection at the population-level, but individual outcomes vary. Achieving natural immunity comes only with infection, and may require several rounds of infection to become strong. On the other hand, vaccination confers a more reliable immunity and appears to lessen risk of reinfection. Both alternatives may involve unpleasant and undesirable effects, but those associated with vaccination are typically fewer, rarer, and less serious than what occurs with natural immunity. Ultimately it seems those who make a choice against vaccines do so motivated by factors other than scientific facts.

Perhaps it will prove to be the case that an anti-vax desire to have effective treatments will coincide with their development so that even if people elect not to get vaccinated there will be effective treatments widely available and at an early stage. An antiviral medication named *molnupiravir* is being developed and

---

[271] Randolph and Barreiro, "Herd Immunity," 729.
[272] Baraniuk, "How Long Does Covid-19 Immunity Last?" 2.
[273] Cho, et al., "Antibody Evolution." (This had not yet been peer-reviewed.)
[274] CDC, "New CDC Study."

tested, as are others (e.g., PF-07321332, and AT-527), with some hope that one or more might be available by late Fall, 2021.[275]

A Cautionary Note

I stated at the beginning of this book my desire to give both sides of the vaccine controversy a fair hearing. To someone simply skimming the contents of this volume, the length of this chapter (and perhaps even its presence) may suggest that the pro-vaccine position is given more attention and is being afforded an unfair advantage. Let me address any such concerns.

First, the specific chapter on anti-vaccine arguments is longer than the one on pro-vaccine arguments. This makes sense in part because there are more arguments listed for the anti-vax position. But it also makes sense because the anti-vaccine position is one *against* an existing position and thus to be fully heard requires some idea of the position being spoken against, which requires a little more space to be gained.

The same thought holds true for this chapter. The rebuttal of anti-vax arguments could be (and too often is) simply a dismissive shake of the head, perhaps with some muttering. I have tried to be more respectful. Though the anti-vax position is a minority one it has been undoubtedly influential and merits serious consideration. Thus this chapter tries to again restate the anti-vax claims (thus adding to its length), then elaborate a full response along pro-vaccine lines, and finally to articulate more completely crucial points being contested along lines pro-vaccine advocates make. Doing all three things simply takes space. It does not mean the pro-vaccine side is being unfairly advantaged.

Second, the next chapter constitutes a continuation of the anti-vax arguments. I have actually divided the anti-vax case into two parts—their claims against vaccines, and what they stand for. The latter provides needed context within which to better understand *why* they make the arguments they do, and also to better comprehend *how* those arguments are shaped and expressed. Looked at in this way it is clear that I have not been one-sided in giving disproportionate space to the pro-vaccine side.

There is a time and a place for assessing the arguments and making judgments about them. I said at the beginning of the book that the time is *after* making a full and fair effort to hear both sides, and the place is at the *end* of such a review. I will indeed provide my assessment (chapter 9) and even go beyond that to offer my own free advice to the parties (chapter 10).

None of these chapters provide coverage of *all* the arguments that might be articulated by those on either side. Neither are all the people who speak out on one side or the other named in these chapters. The aim here has been a much more modest one—to try to represent as fairly as I can the most visible arguments as articulated by prominent people and/or groups on both sides of the debate over COVID-19 vaccines.

---

[275] Aleccia, "A Pill to Treat COVID-19."

Chapter 7
The Anti-Vaccine Positive Position

Anti-vaccine arguments are by their very name and nature *against* a position and thus anti-vaxxers receive much less attention concerning what they are *for*. Outside anti-vaxxer circles what the movement stands *for*—its positive position—is generally obscured by the attention-getting claims against vaccines, mask mandates, and so forth.

So let us turn to what the anti-vax movement is *for*. This returns us to the considerations of the chapter on the heart of the matter for each party involved in staking a position on COVID-19 vaccination. *The positive position of anti-vaxxers advances specific cultural values and ideals.* However, it should be noted that even in promoting certain ideals the language of anti-vaxxers is often couched in negative terms (i.e., as against something).

Freedom as Individual Liberty

Individual liberty—'freedom'—is what the anti-vaccine movement is *for*. Look, for example, at the titles of publications at the prominent anti-vaccine website *Children's Health Defense*. They include *Protecting Individual Rights in the Era of COVID-19* and *Vaccine Mandates: An Erosion of Civil Rights?* When addressing the pandemic the focus is not on vaccination as such (science), but on individual freedom (culture).

This is perhaps clearest in writings of figures like lawyer Robert F. Kennedy, Jr., but it is by no means absent in the writings of anti-vaxxers like pharmacologist Yeadon, who said about government recommendations in the United Kingdom in responses to the pandemic: "I am of the view that the effect of these guidelines approximates censorship."[276] This is a cultural, political stance, not a scientific one. Similarly, as we saw in chapter 2, members of the Association of American Physicians and Surgeons (AAPS) are pushing a cultural agenda hghlighting individual rights of patients and physicians.

For instance, anesthesiologist Paul Martin Kempin, President of the Association of American Physicians and Surgeons (AAPS) during the pandemic, warns, "The COVID crisis has clearly exposed the increasing subjugation of the individual practice of medicine to multiple corporate, media, world, and various government agendas." It is the encroachment of the majority on the rights of individuals that is at the heart of his concern. The AAPS, he notes, stands for the *independent* practice of medicine. In contrast, he writes, COVID-19 mandates have imposed limitations on such practice. To combat this situation, he concludes, requires recognizing that freedom is not free—it requires involvement in the political realm as well as the professional one.[277]

---

[276] Yeadon, "The PCR False Positive Pseudo-Epidemic."
[277] Kempen, "A Perspective."

A notable example of the actual intersection of medicine and politics is seen in the views of physician Joseph Lapado, appointed Surgeon General for the state of Florida in late 2021. In an opinion piece for the *Wall Street Journal*, Lapado wrote, "Liberty has played a special role in U.S. history, fueling advances from independence to emancipation to the fight for equal rights for women and racial minorities. Unfortunately, Covid mania led many policy makers to treat liberty as a nuisance rather than a core American principle."[278]

The Fight to Preserve Freedom

Anesthesiologist Marilyn Singleton sums up a common anti-vax sentiment: "While Americans have been asleep at the wheel, the power brokers were manipulating us into constant societal unrest, with dwindling individual liberties and growing dependency on the government and big corporations. COVID-19 has been a handy justification for left-leaning politicians to promote and solidify their ideology."[279]

This leads endocrinologist David Westbrock to ask, "Do we give up our natural, God-given freedoms and sacrifice the next generation's economic well-being as well as family, social, and religious connections based on statistical prognostications?"[280] Put in other words, which is more valuable—the cultural ideals we have lived by for generations, or scientific projections about what *might* be true?

Insistence on individual liberty produces resistance to *governmental infringements on freedom*. This also can be traced back to the anti-vax movement against smallpox vaccination. Historian Nadja Durbach in her book on the subject observes that, "Alternative medical practitioners consistently claimed that orthodox medicine was a tyrannical system of state-sanctioned interference with the lives and health of an oppressed people."[281] In our current situation Westbrock questions the authority a government can wield on a simple presumption that its 'protection' will do more good than harm. [282]

Today's anti-vaxxers find the threat in a number of different places. Yeadon, as we saw in a previous chapter, fears governments colluding to stratify their populations into those privileged via vaccination and the oppressed unvaccinated creating a new kind of apartheid. He views governments using vaccination status as a way of seeing who gets to go and who gets to stay, whether as citizens or even as living beings.[283]

Marc Marano, author of the book *Green Fraud*, finds a link between COVID-19 "hysteria" and concerns over climate change. He worries that, "A merging of COVID and climate may be just the left's ticket to a Soviet-style regulatory state." He wonders, "Is there a silent majority that will awaken and fight back

---

[278] Lapado, "An American Epidemic of 'Covid Mania.'"
[279] Singleton, "COVID-19," 43.
[280] Westbrock, "COVID-19: Reflections," 54.
[281] Durbach, *Bodily Matters*, 28.
[282] Westbrock, "COVID-19: Reflections," 55.
[283] Delaney, "EXCLUSIVE—Former Pfizer VP."

against climate and COVID hysteria? Are there elected officials in either party ready to lead?" In such comments we see a rhetorical rallying cry, carried out in his appeal to George Washington and the American Revolutionary Army. To preserve liberty, one must resist an oppressive government. At the same time, he accuses government—and the scientists who support its agenda—of having redefined health itself, moving from such traditional ideas as exercise, good diet, and proper sleep, to social isolation, masks, and other pandemic-related measures.[284]

The Right to Make Personal Choices

At the center of what anti-vaxxers mean when they talk about freedom or individual liberty is *the right to make personal choices for themselves*. They fear things like mask mandates deprive them of such choices. The third element in the common mantra of anti-vaxxers is, "It is my body, my choice." It has been adopted by individuals and groups (e.g., Tennessee Coalition for Vaccine Choice and Texans for Medical Freedom) alike.

This is asserting a general right of self-determination, but a particularly strong right with respect to one's own body. The slogan sounds familiar, because it is appropriated from the pro-choice movement on abortion. It has been suggested that the choice of this phrasing is representative of other such choices by anti-vaxxers to adapt previously proven highly emotive things to use in the service of their own movement.[285]

The Tools of Freedom

In pursuit of individual liberty anti-vaxxers utilize a number of traditional tools, including:

- ❖ the exercise of free speech;
- ❖ utilizing a free press;
- ❖ protesting;
- ❖ organizing;
- ❖ political activity; and,
- ❖ legal activity (i.e., lawsuits).

Anti-vaxxers extol the *exercise of free speech*. It is both a key cultural value and a practical tool in the anti-vaccine movement. They often regard themselves as victims of a so-called 'cancel culture.' They see acts like those done by social media giants such as Facebook and Twitter removing their content as prime examples of censorship, disputing the judgment that what they have presented is 'misinformation.' Indeed, in light of the seeming restrictions on free speech they wonder exactly what can be trusted in what is reported. This then gives rise to conspiracy theories, such as seasonal flu cases being coded as COVID-19 rather than what they actually are.[286]

---

[284] Marano, "COVID-19 as a Model." The quotes are from pp. 77 and 81, respectively.
[285] Goyal, Hegele, and Tenen, "Op:Ed."
[286] On all these notions, see Singleton, "COVID-19," 45–46.

Accompanying the right of free speech is the American *right of a free press*. Anti-vaxxers make full use of it in the publication of their views through many mediums, from social media posts, to brochures and pamphlets, and through articles and books. It would be accurate to say that anti-vax publishing constitutes a small industry of its own. But probably the most notable way they exercise this right is through internet websites. It is impossible to accurately estimate the number of such sites because websites come and go, but they number in the hundreds.[287] However, a disproportionally few number of people account for the vast majority of content on social media platforms. On *Facebook*, for example, a mere dozen people account for up to 73% of anti-vaccine content.[288]

A third tool of democracy used by anti-vaxxers is *protesting*. These protests are typically in response to suggestions of mandates (e.g., mask or vaccine), or restrictions on movement (e.g., social distancing) or gatherings (e.g., forbidding gatherings beyond a certain size limit). A common way these protests are conducted is through very public defiance of mandates and restrictions, for instance by gathering in crowds, unmasked. Regrettably, these have generally failed to rise to the bar of being peaceful civil disobedience as the protests are often confrontational in nature, sometimes violent, and typically display an unwillingness to bear the civil consequences of their defiance.

Another tool of freedom is the *right to organize*, and like anti-vaccine movements of old, modern anti-vaxxers organize. We have seen many instances of this in professional and lay organizations alike. Through such organizations and groups the cultural ideals of the movement are shared and identity is reinforced among the like-minded. This is one reason why *social media* has been such a focus. Through it the anti-vax culture can be communicated and grown. An analysis of such websites before the pandemic found they aim at appealing to ideals of choice, personal freedom, and self-determination through informed decisions. They espouse natural treatments and homeopathic remedies as vaccine alternatives.[289]

Many anti-vaxxers are *politically active*. As might be expected given his heritage, Robert F. Kennedy, Jr. is among the best known in this respect. Like many anti-vaxxers, his political activity has been as an *activist* rather than as a politician. In addition to his anti-vaccine activism he also has been a noted environmental activist. Anti-vax entrepreneurs Ty and Charlene Bollinger created a Super PAC named 'United Medical Freedom' before the 2020 elections. The Association of American Physicians and Surgeons (AAPS), is both politically conservative and active, and in 2018 donated only to

---

[287] See, for example, Moran, "Anti-Vaxx Websites," who reports on an investigation of 480 such sites—and that was *before* the pandemic; since then the number has grown.
[288] CCDH, "The Disinformation Dozen," 7. The dozen are listed on p. 6: Joseph Mercola, Robert F. Kennedy, Jr., Ty and Charlene Bollinger, Sherri Tenpenny, Rizza Islam, Rashid Buttar, Erin Elizabeth, Sayer Ji ,Kelly Brogan, Christiane Northrup, Ben Tapper, and Kevin Jenkins.
[289] Moran, "Anti-Vaxx Websites."

Republicans running for federal office.[290] But this is just the tip of a gathering wave that in the last few years has seen a vast rise in founding state political action committees, forming coalitions with other constituencies, and building a vast network to associate with and influence others.[291]

There are also anti-vaxxers who run for political office. For example, Laura Loomer (sickened by COVID-19 in September, 2021) ran for Congress as a GOP candidate in 2020. And the most famous Republican in the country—Donald Trump—has long been associated with anti-vaccine views.[292] But anti-vax candidates have not fared well at the polls, and so many, like Jennifer Larson and Mark Blaxill, who formed a political party in 2011, eventually moved away from directly running candidates to supporting major-party politicians who could be persuaded to sympathize with their cause.[293]

As the above suggests, the anti-vaccine movement in the United States has been strongly associated with the GOP. Indeed, lawyer Teri Kanefield points out that it has been Republican leaders backing political opposition to vaccine and mask mandates, and often vocally supporting anti-vax rhetoric and theories. She points as examples to Governors Kristi Noem of South Dakota and Tate Reeves of Mississippi, as well as Congressman Jim Jordan (Ohio) and Senator Josh Hawley (Missouri). Kanefield ask, "Do some Republicans think they can reap political benefits from the continued spread of Covid? And if so, how might that calculation factor into their policy decisions?"[294] Both anti-vaxxers and pro-vaccine advocates have remarked how often the vaccine controversy apparently has become a political, partisan one. But while rank-and-file anti-vaxxers, with their distrust of government, generally prefer to distance themselves from this element, anti-vax leaders seem to cultivate it.

Finally, *anti-vaxxers use the legal system to advance their policy positions as well as to defend themselves against what they see as unconstitutional oppression.* For instance, Robert F. Kennedy, Jr. and the Children's Health Defense submitted a lawsuit in August, 2020, against Facebook, alleging the social media entity not only was censoring content but engaging in a smear campaign.[295] As a broader example, it is common to see on the news stories of individuals or groups filing legal petitions to challenge mandates related to COVID-19.

In all these ways anti-vaxxers affirm democracy and democratic processes even though they believe themselves generally trodden upon by the majority and their elected officials. One reason the anti-vaccine movement has found so much sympathy among the most conservative Christian evangelicals is that both

---

[290] Khazan, "The Opposite of Socialized Medicine."
[291] Haelle, "This Is the Moment."
[292] Jarry, "The Anti-Vaccine Movement in 2020."
[293] Haelle, "This Is the Moment."
[294] Kanefield, "What Could Be Motivating." She also points out that every GOP Governor has been vaccinated, as well as most Republicans in Congress, meaning, "Republican leaders are getting vaccinated while encouraging others to reject vaccinations, vaccine requirements and mask mandates."
[295] Weir, "How Robert F. Kennedy Jr. Became the Anti-Vaxxer Icon."

groups see themselves as morally sincere groups earnestly trying to preserve the best of traditional cultural values and ideals.

Freedom in a Cultural Matrix

Every culture features especially salient values that serve as aspirational ideals. For anti-vax culture these include:
- ❖ individualism (especially, the individual vs. 'the System');
- ❖ minority status (especially as 'oppressed'); and,
- ❖ naturalistic health (especially 'alternative medicine').

Individualism

Years before the current pandemic, research into anti-vaccine websites, where much of the movement's power resides, found that, "Almost all sites featured the adversarial notion of 'us versus them' whereby parents and antivaccinationists stood against the depersonalized 'them' of doctors, health bodies, governments, and pharmaceutical companies."[296] Among Americans there exists a long history of emotional attraction to the myth of the lonely individual against 'the system'—and this myth fits perfectly with the narrative of individual liberty being threatened by massive, uncaring, unfeeling entities like Big Government and Big Pharma. That is not to say this belief is insincere, nor is it a recent development.

This stance has deep roots. In his history of vaccinations, Blume explains that the anti-vaccine movement that arose in the 1860s against smallpox vaccination (the Anti-Vaccination League) framed itself as an affirmation of certain cultural values, especially respecting individual decisions about one's own body rather than endorsing State-sponsored, legalized 'bodily assault.' These early anti-vaxxers espoused rejection of vaccination as just another life-style choice—like being vegetarian or abstinence from alcohol.[297]

Anti-vaxxers typically believe they are both misunderstood and misrepresented by pro-vaccine advocates. For example, they argue that COVID-19 vaccines should not be *mandated*. Anti-vaxxers typically accept that other Americans are free to accept the vaccine, but believe they should be free to *reject* it. In other words, at a public policy level they are resisting mandates, though at an individual, personal level they are also rejecting vaccines. But this important distinction is often missed by others because they fail to understand the logic of the *cultural individualism* anti-vaxxers embrace.

A culture of individualism views *individual risk* as far more important than *public risk*. Anti-vaxxers emphasize that individual risk from COVID-19 is very low. Since they also embrace an ideology of health through natural means it seems to them quite logical that anyone who lives a basically healthy lifestyle has very little to fear from COVID-19. Thus when they address public risk they emphasize that COVID-19 mostly adversely affects the elderly (with weakened

---

[296] Davies, Chapman, and Leask, "Antivaccination Activists," 23.
[297] Blume, *Immunization*, 32.

immune systems due to normal aging) and those with underlying conditions (especially immune system diseases).

Minority Status

Anti-vaxxers accurately see themselves as holding to a minority position on vaccination and various other traditional values, especially individual liberty in extreme form. We have addressed above a variety of matters in this respect. Here, though, let us focus on how anti-vaxxers navigate life in a democracy.

Though seeing themselves as both a minority and an oppressed one, anti-vaxxers nevertheless do not typically espouse violent overthrow of the government. (Anti-vaxxer involvement in the January 6, 2021 Capitol riots need not necessarily be viewed as espousing insurrection since they may have believed they were defending a legitimate President against a usurper who had stolen the election from him.) Generally they present as American patriots conserving traditional values, upholding both democracy and capitalism.

Anti-vaxxers like Singleton see the pandemic as having become an ally of efforts at extreme social engineering.[298] They generally associate politically progressive or liberal policies as being such social engineering—government sanctioned ways to force unpleasant policies upon a dissenting minority. Thus they gravitate to the GOP, which increasingly presents itself as the party of grievances rather than as one offering a formal agenda of positive action.

What they mourn, as physician Kristen Held writes, comes from what has been lost because of the societal response of the majority to the pandemic: "We have lost our ability to congregate in groups of 10 or more; go to church, school, weddings, funerals, sporting events, concerts; go anywhere without a mask; or hug our parents, grandparents, children, grandchildren, and the lonely."[299] These lost activities are for many too high a price to pay.

Life as a minority can easily lend itself to a rhetoric of being 'oppressed' when social policy decisions run contrary to anti-vax ideals. Such rhetoric plays especially well with conservatives like White Evangelicals who see their historical status and privileges in American society being challenged and eroded by demographic changes producing more and more diversity.[300] Thus there is a degree of overlap between anti-vaxxers and White Evangelicals, though this is sometimes over-exaggerated.

Conservative religious folk may be motivated by genuine belief concerns, as in the fear that being vaccinated might indirectly support abortion, and thus embrace the anti-vaccine movement while also quickly coming to endorse other arguments espoused by it, especially when these are couched in religious terms. Some anti-vaxxers extol 'natural immunity' as properly God-fearing, a choice to honor God rather than trust human technology. If they fall ill, it is God's will.

---

[298] Singleton, "COVID-19," 43.
[299] Held, "COVID-19 Statistics and Facts," 71.
[300] See Bolich, *Two Masters: Evangelicals and the GOP.*

Natural Health

In addition to other ways in which they may see themselves in a minority, they may increasingly view themselves as members of a vanishing traditionalism that espouses natural health rather than dependence on modern medicine. As we have seen throughout this volume, the counter to vaccine use is consistently espousing a reliance on natural health strategies and remedies.

The organization CivicScience (https://civicscience.com) conducted a series of pre-pandemic online polls, with 1,700 to 3,000+ respondents, to get a sense of public attitudes toward vaccines, particularly childhood vaccines. These polls do not claim themselves to be representative samples of the American population. Poll results are achieved by gathering data from voluntary participation as people visit various websites partnered with CivicScience. Among other matters, the collected data revealed that those who favor an anti-vaccine position are more likely young (under 30) and are far more likely to support natural health values like being concerned about GMOs (genetically modified organisms) in food and purchasing organic food.[301]

A large part of the justification for being anti-vaccine relies on the safety and efficacy of alternatives to vaccination. In that regard the first line of defense is touted as natural health strategies like proper diet, sleep, and exercise. These function as preventive in nature. To them are added things like health supplements to play a role not only in prevention, but also in treatment.

The Bottom Line

What all these variables share most in common is *a concern for cultural values over a culture of science.* Pro-vaccine advocates seldom, if ever, talk about their advocacy of a 'culture of science.' Yet such a culture is part of the American fabric, particularly as expressed in practical science through avenues such as technology and medicine. The pragmatic test of truth—a thing is true if it works—accepts a relative lack of knowledge in the interest of getting things done. It is easy to criticize that as a willingness to cut corners or undervalue safety, both of which sometimes occur, but it is a general spirit behind what most Americans see as why the country has led the world, including such 'giant leaps' as putting a man on the moon. While pro-vaccine advocates are prone to see the rapid development and distribution of the COVID-19 vaccines as another such 'giant leap,' anti-vaxxers are more pessimistic. Because so many other things are culturally more important to them than practical science, they view with suspicion the enthusiasm and optimism of pro-vaccine proponents.

---

[301] Commisso, "Vaccine Hesitancy."

Who Are Anti-Vaxxers?

Who are the people who embrace this culture and on its value base reject vaccines? Who are the anti-vaxxers? Surprisingly, it seems a question too seldom asked. Pro-vaccine advocates are so focused on the *science* aspects of the vaccine controversy they often forget to look at the *people* aspects. But trying to understand who are the people most likely to embrace anti-vax culture necessitates an effort to *see* them.

In social science, we are interested in both individuals and groups. In trying to form a picture of any group one basic procedure is to discern the group's demographics. Of course, in any group there are likely to be individuals who do not at all look like the demographic picture. And there will be some individuals who reflect particular demographic characteristics and not others. So any demographic picture must be handled gingerly by avoiding drawing overgeneralizations or trying to apply a general description to every specific member of the group.

With those caveats in mind, let us examine what some research has found. First, we might consider a broader designation, like 'vaccine hesitant,' that may include anti-vaxxers but also embraces many not willing to identify themselves that way but who for one reason or another have not accepted vaccination. Thus a May, 2021, report sponsored by the U.S. Department of Health and Human Services through the Assistant Secretary for Planning and Evaluation (ASPE) gathered together both those would said they would probably not get a vaccine (i.e., 'hesitant') and those who said they would definitely not get a vaccine (i.e., 'strongly hesitant') under one large umbrella. For the period March 17–29, the data revealed the following about the 'vaccine hesitant':

- they are as likely to be female as male;
- they are more likely to be young (22% ages 25–39; 18% for both ages 18–24 and 40–54);
- they are slightly more like to be Black (18%) than White (16%);
- they are more likely to have no college education;
- they are more likely to be found in the South, Great Plains, and Alaska than in the West or Northeast. [302]

Similarly, a Kaiser Family Foundation study in 2021 that compared the vaccinated to the unvaccinated found that the unvaccinated were younger (70% under age 50), less educated (46% high school or less), poorer (42% making less than $40,000/yr.), more likely to identify as Republican (49%), and uninsured (24%). Though 56% of the unvaccinated were White, compared to the overall group Whites were more likely to be vaccinated, while both Blacks and Hispanics were less likely to be vaccinated.[303]

Finally, a study published in *Social Science and Medicine* reports that among the nearly one-third of Americans expressing an intention not to pursue COVID-19

---

[302] Beleche et al., "COVID-19 Vaccine Hesitancy."
[303] Sparks, Kirzinger, and Brodie, "KFF COVID-19 Vaccine Monitor."

vaccination, there were more women, conservatives, and Black Americans. Women, compared to men, were more likely to refuse vaccination over concerns about its safety or efficacy. Blacks, compared to Whites, were more likely to offer every available reason (not safe, not effective, no insurance, no financial resources, already had COVID-19). Being conservative made it more likely to offer as a reason concerns over safety, efficacy, or insurance.[304]

While other research might be mentioned, the above are sufficient to indicate in broad strokes the kind of characteristics associated with people who to one degree or another identify as anti-vax or are sympathetic to its position.

An Important Caveat

Everything in this chapter so far has attempted to cast the anti-vaccine movement in as positive a light as possible, attributing to them genuine sincerity of convictions. But there are some signals within the movement that ought not to be ignored. Specifically, it may be possible to distinguish between the motivations of the small group who constitute the ideological leadership of the anti-vaccine movement—what we might term the 'professional anti-vaxxers'—from the multitude of followers who have 'bought' the ideological package the leaders advance. The former group, really a very small number of folk who make much if not most of their income profiting from being anti-vax, may not be as invested in the anti-vax ideology espoused above as are their followers.

Various researchers studying the history of the anti-vax movement have suggested that in recent years there has been a calculated rebranding by anti-vax leaders in order to broaden their appeal. This creates the worrisome possibility that anti-vax leadership is operating less by conviction than political calculus. For example, science journalist Tara Haelle, author of *Vaccination Investigation: The History and Science of Vaccines* (2018), notes in a 2021 article that in the past half-dozen years they have organized politically like never before. She argues, "Their versatility and ability to read and assimilate the language and culture of different social groups have been key to their success." Haelle refers to the work done by Stanford researcher Renée DiResta, who in analyzing anti-vax messaging finds a conscientious move to adopt the vocabulary most likely to resonate with other conservative groups, focusing on 'freedom' and 'choice.' In so doing anti-vax leadership has been successful in enlisting the support of conservative groups and politicians.[305]

The Bottom Line

I imagine the overwhelming majority of Americans, no matter which side of the vaccine debate they are on, prize individual liberty. It is a widely shared cultural value. But it is also one that brings us to *ethics* as a possibly fertile meeting ground for profitable dialog rather than mere shouting across a cultural divide.

---

[304] Callaghan et al., "Correlates and Disparities of Intention."
[305] Haelle, "This Is the Moment."

Chapter 8
An Intersection at Ethics

From what we have seen so far, it might seem difficult at best—and impossible at worst—to think the opposing sides might be able to find common ground. But let's be optimistic. Perhaps an intersection of viewpoints can be found in *ethics*—a concern both in the culture of science and in the anti-vax culture, whether in the sphere of religion or politics.

Ethics structure moral decisions and behaviors. Ethics help coordinate morality so that people make consistent choices about what is right and wrong. We see that happening with respect to COVID-19 pandemic behavior. Those who are pro-vaccine and have become vaccinated are also more likely to continue to stay away from large groups, wear face masks when around people outside their home, and avoid nonessential travel.[306] Those who are anti-vaccine are more likely to act otherwise, and oppose mandates in public fashion.

Among anti-vaxxers those who identify as White Evangelicals may seem especially receptive to ethical considerations. After all, ethics and moral behavior are key aspects of Christianity. A definite attitude about social responsibility—caring for the health and well-being of others—is evident in both the Hebrew Bible and the New Testament. But whether one is a conservative Christian or not, thinking about ethics and morality in the United States has been profoundly influenced by Christianity and so we profit by beginning there.

'Love Your Neighbor as Yourself'

In the Hebrew Bible—the Christian Old Testament—the question arises very early when Cain asks God, "Am I my brother's keeper?" (Genesis 4:9). The remainder of the Bible can be read largely as an answer in the affirmative to that question. For example, a good portion of biblical law concerns how to live with others in both private and public spheres.

In the New Testament, Jesus answers the question as to what is the greatest commandment this way (Mark 12:29–31 (NIV)):

> [29] "The most important one," answered Jesus, "is this: 'Hear, O Israel: The Lord our God, the Lord is one. [30] Love the Lord your God with all your heart and with all your soul and with all your mind and with all your strength.' [31] The second is this: 'Love your neighbor as yourself.' There is no commandment greater than these."

The idea of loving one's neighbor has been a cornerstone of Christian ethics. The notion has been fundamental to the American sensibility, too. In fact, with specific reference to COVID-19 vaccination, *90%* of American adults

---

[306] AP-NORC Poll (Aug. 20, 2021).

say health effects for the community should be part of making a personal decision.[307]

It is surprising then to find that Pew Research Center polling found White Evangelicals are less inclined than members of any other American religious group to think community health concerns should be an important factor in their deciding whether or not to be vaccinated.[308] Perhaps some of this is to be attributed to a Protestant legacy stemming from the fiery independence of Martin Luther, and perhaps part can be attributed to the American culture in which White Evangelicalism has its roots, and perhaps part can be assigned to the willingness of White Evangelicals to subordinate their religious ideology to the political ideology of the GOP. Whatever the sources, it is a stance at variance with the majority of Christendom and Christian history.

So let's focus on the "as yourself" part of Jesus' statement. Can, or should 'love of yourself' include vaccination?

Earlier in this volume we saw that while all American religious groups favor getting the vaccine, the group least likely to endorse that choice is White Evangelicals. Still, among them too exist deep divisions, with some siding with the position that science should decide the question and others resisting vaccination based on the kind of cultural considerations covered in the previous chapter. They do seem to agree that the choice should reflect cultural values associated with their *beliefs*. That provides a useful base upon which to build.

Well-respected and trusted organizations like the National Association of Evangelicals (NAE) have made efforts to address anti-vax Christian concerns on specifically religious grounds (without giving up science). Like anti-vaxxers they choose to focus on cultural—in this case, religious—concerns and values. The "Christians and the Vaccine" project, spearheaded by the NEA and the Babel Project, states, "Our goal is to equip pastors and Christian leaders to help others apply biblical principles to this topic. Based on these principles, we encourage Christians to take the vaccine."[309]

One notable feature of the "Christians and the Vaccine" project is their explicit reliance on Jesus' teaching. They write:

> [W]e believe that the core truths of Christian teaching support taking the vaccine. Chief among these is Jesus' 2nd greatest commandment to "love your neighbor as yourself." The very nature of vaccination is "public" health, meaning it is just as much about your neighbor as about yourself. We have successfully tackled polio, small pox, measles and other deadly diseases through the tool of vaccines. The COVID vaccine is no different.[310]

Similarly, the Bible-believing Christian nonprofit BioLogos offers reasoning and information as to "why we think vaccination is a safe, ethical, and wise

---

[307] Funk and Tyson, "Growing Share of Americans," 28.
[308] Funk and Gramlich, "10 Facts about Americans and Coronavirus Vaccines."
[309] Christians and the Vaccine, "About Us."
[310] Christians and the Vaccine, "FAQs."

choice for Christians." In other words, like the just-mentioned project, they address both the science of immunization as well as common anti-vaccination concerns rooted in culture. Importantly, they specifically tackle the cultural issue of ethical concerns about possible use of fetal cell tissue in vaccines (addressed in a previous chapter). They argue, "While the association with abortion gives many Christians pause, there is substantial agreement among Christian theologians and ethicists that the connection to fetal cell lines should not make these vaccines off-limits for Christians."[311] (And remember, vaccines like Pfizer's and Moderna's do *not* have any connection to such cell lines.)

BioLogos also appeals to the New Testament, circling back to the ethical thinking that involves care and concern for others. In this case they look to the words of the Apostle Paul. They cite Philippians 2:5–7 (NIV):

> [5] In your relationships with one another, have the same mindset as Christ Jesus:
>
> [6] Who, being in very nature God,
> did not consider equality with God something to
> be used to his own advantage;
> [7] rather, he made himself nothing
> by taking the very nature of a servant,
> being made in human likeness.

In this context it is not a matter so much as following the *command* of God to love others, but instead choosing to follow the *example* of Jesus who humbly subordinated his own 'rights' as God to the position of being a servant.

We might also think of Paul's letter to the church at Rome. In it, in the latter chapters where he focuses on ethical issues, he reminds his hearers, "For none of us lives for ourselves alone, and none of us dies for himself alone" (Romans 14:7 (NIV)). The inescapable truth is that human beings are communal beings and what we do affects others.

The Ethics of Freedom

So what do such texts and the thinking behind them mean when we turn to the grand concern of anti-vaxxers for personal freedom of choice? To get an answer we need to start very simply.

Let us begin by recognizing that Americans on every side of the question of vaccination value freedom, cling to individual liberty, and uphold the right of personal choice. It is how these are applied in living together that generates differences. What we need, then, are some basic principles, or axioms, that might appeal to all of us.

Axiom 1: Freedom Has Limits

A basic axiom in the ethics of freedom is, 'Your freedom ends where my nose begins.' Put more formally, *personal freedom has limits, or borders*. Both pro-vaccine people and anti-vaxxers can agree to this principle. The pro-vaccine person says, 'You can choose to go unvaccinated and without a mask, but that

---

[311] BioLogos, "Should Christians Get Vaccinated."

doesn't confer on you the freedom to interrupt my liberty to do both.' The anti-vaxxer turns that around to say, 'You can choose to vaccinate and wear a mask, but you can't interfere with my right not to!"

How can such views be reconciled? Perhaps a better question is, Do they need to be? The most practical solution is *self-segregation*—something people do anyway, whether it is always a good idea or not. As long as anti-vaxxers separate themselves from the vaccinated they pose minimal risk. But are they willing to do so? Human nature being communal we commonly create many public spaces in which we do commerce, socially interact, and carry on much of the business of everyday living. So self-segregation is not a sufficient answer.

Both anti-vaxxers and pro-vaccine folk get that personal liberty has borders. Even physicians sympathetic to the anti-vaccine movement, like Jane Orient, head of the Association of American Physicians and Surgeons (AAPS), acknowledge such limitations. Orient points out that an unvaccinated person with no exposure to a disease should not have her or his freedom restricted, but that potentially contagious individuals can rightly have their freedom of movement restricted in the interest of the safety of other individuals.[312] Similarly, Deborah Daniels, former assistant U.S. attorney general, and president of the Sagamore Institute, observes that from a legal standpoint, "A person's constitutional freedoms don't apply when his actions are harming others. And vaccine mandates in the interest of public health and safety were determined to be constitutional over a century ago by the U.S. Supreme Court."[313]

Let's all be honest. Americans—whether anti-vax or pro-vaccine—accept restrictions on personal liberty regularly. Every time we strap on a seat belt, stay in our lane while driving, or stop at a red light we are accepting a restriction on personal liberty. We are choosing to obey laws and policies established to protect public safety, and we are accepting the underlying notion that the 'public' is a collection of individuals who agree mutually to constrain our individual choices and exercise of liberty.

Axiom 2: Freedom Must Be Equal

The second axiom posits that both sides are right in affirming their individual liberty. But how do we live together if only one side is allowed to exercise that freedom? It would seem another old saying offers a second basic axiom: 'What is sauce for the goose is sauce for the gander.' This is the principle of *equality* in the application of freedom. It simply means that a freedom granted to one side pertains as much to the other. Likewise, what constrains freedom on one side constrains it equally on the other.

While initially this may seem especially burdensome on the pro-vaccine side—after all, they are in the majority!—the insistence that equality of liberty be preserved for all is quintessentially American. The underlying logic is that *no one's liberty is safe unless everyone's freedom is protected*. Much like Ben Franklin

---

[312] Orient, "Statement."
[313] "Vaccines Are Not Just a Matter of Personal Choice."

declaring that if we do not hang together we shall certainly all hang separately, every individual has a vested self-interest in protecting the liberty of her or his neighbor.

But neither axiom exists alone. In fact, they are inseparably linked.

Axioms 1 and 2 Operating Together

Each side has an equal right to cling to its decisions, but only as far as adhering to the first axiom. That means *everyone* has a right to make choices about the vaccine, but *no one* has a right to do so in a manner that deprives another of the same freedom in equal measure.

This is why Juliette Kayyem, a former assistant secretary for homeland security and someone recognized as an expert on national security matters, puts it this way: "Americans are entitled to make their own decisions, but their employers, health insurers, and fellow citizens are not required to accommodate them. . . ."[314] Liberty flows both ways—not just in the direction one side desires. Every personal choice carries consequences.

Being free to choose a course of action is not the same as being guaranteed the consequences one desires from that choice. If, for instance, I choose to drink and drive on the freeway as an exercise of personal liberty, it in no way infringes on the right of others to punish me because my choice has violated the basic axiom of threatening their safety. This sensibility undergirds much of law.

The American philosopher John Rawls writes about a *social contract* we all enter into.[315] This isn't a literal, written contract, but it is powerful nonetheless. It is, for example, in operation when we pass one another on the road going in opposite directions. We both presume the other will stay in his or her lane. We presuppose we are equal and free, but we also accept that we exist in a situation where successfully living together requires agreeing upon certain principles as to what is just in our interactions.

A consequence of accepting that we are both free and equal—that both axioms presented above are in effect—is that *we willingly draw borders to our personal freedom in order to preserve the equality of freedom of others.* No one's freedom is safe unless everyone's freedom is secured.

What has unfortunately transpired in confrontations between pro-vaccine and anti-vaccine advocates is that each side extols liberty, but not equality. Thus a pro-vaccine group might argue that as a majority in a democracy they have an unequal right to see their ideas of how liberty should play out. The logic amounts to, 'We won the popular vote so we should get our way even if it infringes on others' rights.' On the other side, anti-vaxxers may argue that the crushing oppression of the majority cedes the moral high ground and thus puts anti-vaxxers in the right such that they are unequally free to infringe on pro-vaccine people's liberty (sometimes in a literal 'in-your-face' manner).

---

[314] Kayyem, "Vacine Refusers."
[315] Rawls, *A Theory of Justice.*

In practical terms, where does this leave us in public contexts where social interactions occur, like at stores, concerts, games, or schools? If, for example, the anti-vaccinated choose to shop unmasked, is that inherently worse than the vaccinated preventing them from doing so? Is it inevitable that one side must give way to the other, and that then means the oppression of the minority by the majority?

Some pro-vaccine folk would argue further that in states of public emergency the borders of freedom are modified in the interest of the entire community. This is the ethics behind *public mandates*. Kayyem uses the example of a sinking ship, pointing out that while individual passengers may choose not to wear a life-vest the circumstances don't allow taking a lot of time to decide whether to do so or not. As she observes, emergencies may narrow options down to some alternatives that under more ideal circumstances would never be considered. Just as with a sinking ship, where telling people to put on a life-vest seems reasonable, she argues in a pandemic asking people to get vaccinated is not much of an imposition—*in that context*.[316]

As we have seen throughout this book, context matters. Consider, for example, the *Joint Statement in Support of COVID-19 Vaccine Mandates for All Workers in Health and Long-Term Care* (July 26, 2021). Signed by more than 50 health-related organizations, the Statement's basic argument is a simple one: "Vaccination is the primary way to put the pandemic behind us and avoid the return of stringent public health measures."[317]

Note the final clause--*and avoid the return of stringent public health measures*. In other words, a mandate may not be an ideal, but it is the practical alternative to *more stringent* measures. Put bluntly, *restricting individual freedom in some limited context facilitates preserving individual freedom in wider contexts*. Thus anti-vaxxers can morally reconcile themselves—if they choose to do so—to accepting the undesirable choice of vaccination as 'the lesser evil' to imposed restrictions such as those that occupied most of 2020.

Can such a perspective be applied in a manner that respects the personal liberty, power of choice, and equal rights of anti-vaxxers? The simplest solution is to grant anti-vaxxers full and equal freedom to choose not to be vaccinated *so long as* they accept the consequences of the equal right of the vaccinated to not be adversely affected by their choice. This might mean, for example, anti-vaxxers who choose to defy vaccine mandates willingly accepting the consequence of losing their job because of their employer's equally free choice. Remember, the choice is painful, too, to the employer who loses workers.

Of course, anti-vaxxers are likely to see this as an unequal application of the axioms. This brings us inevitably to reconsidering the matter of *personal choice*. With the idea of consequences in mind, how can we reasonably and ethically determine moral culpability (i.e., responsibility as being in the right or the

---

[316] Kayyem, "Vacine Refusers."
[317] See American Public Health Association (APHA), "Joint Statement."

wrong) when the personal choice of one individual adversely affects another person? Philosopher Peter Vallentyne proposes that a person is not morally culpable for what results from a personal choice *if* the person could not have changed the outcome by choosing differently, *or* the person could only have avoided the outcome at an unreasonably large cost.[318]

Unfortunately, here we probably have a situation where pro-vaccine advocates emphasize the first criterion, while anti-vaxxers emphasize the second. In other words, pro-vaccine folk will argue that an unvaccinated person who spreads the infection is morally responsible for others getting sick because he or she could have chosen vaccination and dramatically reduced the likelihood of that outcome. But anti-vaxxers are likely to contend that choosing differently comes at an unreasonably large cost because it asks them to foreswear things more dear to them, things at least as valuable to life as lowering risk of infection by COVID-19. Besides, they are likely to add, they believe the risk of infection is the same for both the vaccinated and unvaccinated.[319]

If we cast the matter in a different light, moving away from what may seem abstract philosophical reasoning to the more concrete realm of human experience, perhaps we can again achieve some common ground. Consider a parent with small children strapped in car seats who stops on a hot summer's afternoon to dash inside a convenience store—only 'for a moment'—and leaves the children alone in an unventilated vehicle. The children subsequently die of heat exhaustion. Is their death attributable to the parental choice? American law thinks so—and Americans, whether anti-vax or pro-vaccine, are likely to agree. On the basis of the first criterion set out above, it is clear the parent could have altered the outcome by altering the choice, so there is moral responsibility for the choice. The parent acted wrongly. Both sides are also likely to agree that the second criterion applies, because what cost is unreasonably larger than one's children's lives?

And that realization brings us to an important insight: *how do we measure the cost of a personal choice if not by something as valuable as human life?*

Axiom 3: Responsibility to Safeguard the Preeminence of Life

Freedom is chaos without acceptance of *responsibility*. Every personal choice carries consequences and morally we all must consider not merely the consequences for ourselves, but for others, most especially those others who depend upon us (e.g., family and the vulnerable of our communities). I doubt

---

[318] I am following Cappelan et al., "Choice and Personal Responsibility," in their presentation of Vallentyne, "Brute Luck and Responsibility."

[319] It is demonstrably *not* the same risk. According to one study across 13 states, the unvaccinated are 4-5 times more likely to get infected, 10 times more likely to require hospitalization, and 11 times more likely to die; over 90% of hospitalized patients with COVID-19 are unvaccinated (Santucci, "Unvaccinated"). However, as a September, 2021 Gallup poll found, unvaccinated Republicans erroneously believe the rate of infection to be the same, and very low (5%) (see Blake, "How Badly Unvaccinated Republicans Are Misinformed," following Rothwell and Witters, "U.S. Adults' Estimates of COVID-19 Hospitalization Risk.").

there are many, either anti-vax or pro-vaccine, who dispute this notion. We *are* our brother's keeper to some extent.

So let us introduce a third axiom: *neither liberty nor equality have power without life and we are all responsible to safeguard it.* I am mindful of that famous and stirring speech of Nathan Hale that concluded with the ringing declaration, "Give me liberty, or give me death!" There are things worth dying for, and personal liberty might be accorded among them. But I'd like to point out that Hale fully intended to go on living. He was not making a suicide speech. He, like today's anti-vaxxer—*and* pro-vaccine American—wanted a nation free from oppression. He wanted liberty so as to *live*.

Anti-vaxxers have every right to echo Hale's words. They have every right to work through political means to change government in the direction of being, as they see it, less oppressive. But they don't have the right to risk the lives of others any more than do pro-vaccine people.

Both sides, morally speaking, must *reckon the cost of personal choices*. The modest proposal here is that the sanctity of life be agreed upon as the criterion. It is not especially difficult for the vaccinated to accept the consequences of their choice for others because their choice protects those others *without infringing on the equal liberty and right of the unvaccinated to continue risking infection*. On the other hand, because the unvaccinated have so much higher a risk of infection and spreading it, unless they voluntarily segregate away from others they *inherently pose a risk to others both vaccinated and unvaccinated*.

The essential difference between the sides resides in the *stakes* when our third axiom is applied. Anti-vaxxers by their choice accept the risk of natural infection for themselves, and are staking their lives on that choice not being fatal. But *they do not have the moral right to accept that risk on behalf of others*. An unvaccinated parent who falls ill and dies abandons children who live to bear the consequences of a choice that both could have been otherwise (Vallentyne's criterion 1), and that could have been avoided without such a high cost (criterion 2). Even worse is the prospect of surviving an infection that sickens or kills those to whom the person spread the infection by their choice.

Accordingly, pro-vaccine people do have a right to insist on their protection through policies like mask mandates, social distancing, and proof of vaccination in places where people gather because the stakes—literally life or death in potentiality—are so high. It isn't an infringement of the liberty or equal rights of anti-vaxxers *in a social context* because the axiom of prioritizing life is just and fair. It is just because we all do live with a social contract that presumes where death or serious injury exists as a real possibility that *everyone* will curb their inherent freedom by choosing to protect life.

Making moral choices where one feels compelled to do what they would rather not do may be uncomfortable and unpleasant, but it is also *responsible*. It is, in fact, what anyone would want of another in one situation or another. That is the very logic behind the doctrine of negligence—expecting another person to do what is reasonable and usual in any given situation. Let's be responsible.

Chapter 9
The Bottom Line

This little book was finished at the end of September, 2021. I am open to the likelihood that the COVID-19 pandemic will continue to be a battleground with shifting views on any number of issues. But as of this moment, the data from different polling sources suggest that American attitudes toward vaccination continue to shift in the direction of favoring it. Put simply, the American public has joined the scientific consensus in declaring in favor of vaccination.

Decline in Anti-Vaccination Support

The Axios-Ipsos Coronavirus Index of August 31, 2021, for example, found fewer Americans stood in hard opposition to getting the vaccine than at any point since the index began. At that point, only 20% of Americans indicated they were not likely to get the vaccine and only 14% (1-in-7 people) were in hard opposition to it. In fact, even on such sensitive matters as employer-mandated vaccination, more than half (57%) of Americans supported it.[320]

Is this a positive trend? Every reader can decide an answer, but it suggests that *proponents of vaccines have succeeded in making a better case than anti-vaxxers*. On the other hand, though dealt a losing hand, *anti-vaxxers have proven successful enough at bluffing as to sow seeds of doubt in many more minds than they have actually have convinced*. For centuries it has been a common ploy when faced with an opponent that has better arguments to shift the battleground as much as possible away from inconvenient facts to more malleable emotions. Thus it is understandable that anti-vaxxers who can't convince people based on actual reason and evidence prefer to use fear-mongering, capitalizing as much as possible on the anxiety engendered by the pandemic.

It may be fair to say that lessened opposition to vaccination has been a gradual trend due to at least two factors: the stark harsh reality of the continuing pandemic and a willingness in light of it to reconsider the anti-vax arguments. The latter, under scrutiny, simply have not fared well.

Assessment of Anti-Vaccine Arguments

Why? Anti-vaccine arguments are often little more than *claims*. At the crassest level—one found commonly in casual conversation—the claims boil down to a well-rehearsed three:

1. "I have done research."
2. "I have expert authorities on my side."
3. "It's my body, my choice."

---

[320] Ipsos, "The Wall of Vaccine Opposition." Also see Beleche et al., "COVID-19 Vaccine Hesitancy."

As this list indicates, a proposition is set forth, frequently with no support. When support is offered it is commonly sparse and vague, with name-dropping of anti-vax websites, or supposed authorities (claims #1 and #2). When more extensive comments are made they typically shift to the core declaration of anti-vax cultural values—personal freedom (claim #3). The last claim often receives better elaboration or offers more clarity because it is what the anti-vaccine movement is most about, at least for those sincere in their convictions rather than cynically exploiting the sincere for personal profit. When pressed, anti-vaxxers seem to inevitably default to claims like the third one listed above.

It is worthwhile in assessing the deficits of anti-vaccine arguments to look more closely at each of these common assertions.

"I have done research."

Doing research is laudable; I recommend it highly, including 'looking again' ('re' (again) + 'search') at the sources used by this book. However, most people on both sides of the controversy tend to look at only a few things and those things they look at are usually what already support their thinking. Anti-vax arguments strongly rely on unsupported claims made over-and-over, as if sheer repetition of a claim makes it true. An analysis of 480 anti-vax websites—before the pandemic began—found, as the researchers put it, "They play fast and loose with the facts (that is, if facts are even employed)." The study found the following:

- ❖ 67% of the sites offer pseudo-science as support for claims;
- ❖ 59% appeal to 'authorities' whose actual expertise is not supported;
- ❖ nearly a third rely on emotional appeals through anecdotes.[321]

Before going any further, let's make one point clear: there may be good science and bad science, but good science is not established by *liking* it. Bad 'science' is typically pseudo-science, meaning it adopts the semblance of science without its actual substance. Anybody can copy scientific terminology and fiddle with statistics to make fantastic claims. In fact, at least as far back as Aristotle there was a recognition that people tend to believe outrageous claims because they can't fathom anyone would tell such a whopper without there being at least some grain of truth to it.[322] When people have a poor case they often resort to shouting louder, repeating themselves, and doubling down by making their claims ever more exaggerated—all to at least sow a little *doubt*. Maybe, the person thinks, the anti-vaxxer might *possibly* be right. And that belief in the possibility the anti-vaxxers *might* be right is enough to persuade many people to at least hesitate.

The more persuasive anti-vax proponents and anti-vax presentations use a particular tactic to reassure those listening to them that they are presenting good research. They themselves urge people 'Do your research,' presumably meaning

---

[321] Moran, "Anti-Vaxx Websites." For the full report, see Moran et al. "Why Are Anti-Vaccine Messages so Persuasive?"
[322] See Aristotle, *Rhetoric,* 1400a [II.23.22].

both that they have done theirs and that if you check them out their claims will hold up. That's good advice, and emotionally reassuring, but it seems anti-vaxxer authors may not actually think people will do so.

For example, anti-vax arguments commonly appeal to scientific studies in respected journals as presenting research that substantiates their claims. But as far as I can tell, either anti-vax authors don't understand what they are reading, or they are relying on their readers not going to their sources (or if they do, understanding what is there)—or both. Actually following up on the sources they cite shows anti-vaxxers persistently misunderstand and/or misrepresent the scientific literature. A good portion of the chapter on the rebuttal of anti-vax arguments consists in showing—again, and again, and *again*—that scientific literature is being misused. Whether that stems from the anti-vaxxers themselves not understanding what they are reading (a charitable explanation), or intentionally misrepresenting it for deceptive purposes (an alternate explanation) is not something I can decide, but either way their original advice should be taken: do your research, i.e., *check out what the sources cited are actually saying.*

"I have expert authorities on my side."

Anyone trained in research can quickly see the ways in which anti-vaxxers frame their arguments: they maximize emotional appeals (specifically arousing anxiety and fear), and try to undergird their contentions by appeals to 'expert authorities.' Let's focus on that latter point. The idea that an anti-vaxxer has 'expert authorities' on their side tends to arise from two ways anti-vax arguments are commonly presented:

❖ the anti-vax presentation (e.g., article or website) is endowed with an impressive title, and/or
❖ the anti-vaxxer authorities are labeled 'expert' and 'authoritative' based on an academic credential (e.g., PhD or MD) and/or experience (e.g., 'pharmaceutical industry insider').

First, presentations of anti-vax claims are often headed by reassuring titles such as "SARS CoV-2 Virus and COVID-19 Vaccine Information." It *looks* scientific and authoritative, inviting trust at once. Anti-vax websites invite people in with such authoritative sounding names as "National Vaccine Information Center (NVIC)," "Urban Global Health Alliance," or "Organic Consumers Association (OCA)." Some utilize built-in sympathy markers with key words like 'children' ("Children's Health Defense (CHD)") or 'informed consent' ("Informed Consent Action Network (ICAN)")—buzz words that the presenters know everyone is in favor of and hold warm endorsing thoughts about. Anti-vax websites often don't openly declare that they are anti-vax. They rely on clever branding to draw people in. The old saying, 'You can't judge a book by its cover' is apt here. So, too, is the even older caution, *caveat emptor* ('Buyer, beware!').

Second, most people are accustomed to accepting that people with advanced educational degrees must have earned them by being smart and knowing things ordinary folk don't. Anti-vax platforms are full of quotes from—and references to—people with PhD degrees. But very frequently (and as indicated here and there throughout this book) these advanced degrees are *not* in the area for which they are being touted as authoritative experts. Just as being an expert with car transmissions doesn't guarantee a person knows how to fix a carburetor, so being a physician credentialed in dermatology doesn't mean the person understands epidemiology. But anti-vax arguments typically eschew such distinctions and treat every PhD or MD on their side equally as 'experts' to whom the rest of us should listen and accept every word as true.

Unfortunately, there don't seem to be enough anti-vaxxers with a PhD or MD to go around, so often enough other credentials are advertised. For example, the charismatic Brandy Vaughan—touted as a 'Joan of Arc' figure[323]—presents herself on her Learn the Risk website as "a former pharmaceutical insider."[324] This sounds more impressive than the mundane reality—for three years she was a pharmaceutical representative for Merck trying to sell Vioxx (an arthritis drug). Exactly how this experience qualified her to speak expertly on vaccines is left unaddressed. Look closely at all claims of authoritative expertise.

Put bluntly, often an illusion of expert authority is being cast by anti-vaxxers. Whether anti-vax or pro-vaccine, what constitutes an 'expert authority' should be considered carefully and wisely. To use myself as an example, I hold two earned PhDs, but neither is in dentistry and so I wouldn't suggest anyone take what I say about dental care as coming from an 'expert.' For that matter, I am also not an epidemiologist, pharmacologist, or expert in genetics. So I should not be appealed to as an 'expert authority' based on my degrees. Any credibility I have comes simply from an accurate presentation of established facts well-reasoned into sensible explanations. In short, 'authority' should be a matter not of *person* but of *presentation*. If the latter meets the criteria of being fact-based, an honest account well-reasoned, then that should be judged more authoritative than a personal appeal such as, "I'm a doctor, I know what I'm talking about, and you should believe me because *I* say so."

That doesn't mean what anti-vaxxers say is automatically wrong, and in fact one could argue that having earned a doctorate presumes being trained well enough in research so as to be better at it than most folk. But unless there is transparency about the credentials, take any claims of expert authority based on holding a degree with a grain of salt. Instead really look at the arguments themselves and test them.

"It's my body, my choice!"

The third claim is undoubtedly the most often heard: "It is my body, my choice!" The claim is appealing, the logic so simple and persuasive that it seems irrefutable. Leaving aside for the moment the irony of this slogan being appropriated and repurposed from the abortion battles of this country, let us

---

[323] Welsh, "Santa Barbara Coroner." The Joan of Arc label seems to some apt as her unexpected death in 2020 left some anti-vaxxers promoting the notion she was murdered by Big Pharma.
[324] See "Who We Are" (https://learntherisk.org/).

begin by asking a simple question: "Is it an absolute right to do what one wants with one's body?'

The obvious answer is *No*. The law, religious teachings and moral philosophers all concur that one does not have a *right* to kill oneself. *Self-abuse* is universally looked down upon. For examples, intentionally cutting oneself with a razor blade, or choosing to eat foods known by the person to be harmful, may indeed be choices, but hardly defensible ones. So there is no absolute right to choose whatever one wants because it is 'my body, my choice.'

Still, the anti-vaxxer isn't arguing for a right to do something harmful *to* the body, but claiming they are standing for a right to *not* do something to protect the body. That is certainly more defensible logic, though it runs into those inconvenient facts to which we keep directing attention. The evidence is in and it's indisputable: an unvaccinated person is significantly more likely to become infected, to be hospitalized, and to die. More than 90% of those in hospitals across the United States right now are unvaccinated.[325] This is not an isolated finding but a consistent one across studies.

An erroneous belief is still a wrong one no matter how sincere it has been formed and is being held. To use an example: when you step off a cliff, gravity does not care if you *believe* you can fly. In the light of the facts, the *passive* act of refusing vaccination constitutes an *active* risk that poses the prospect of harm to both self and others. As we saw last chapter, that is a morally indefensible position to take. One does not have a right to make choices for oneself that can adversely affect the health of others.

Again the anti-vaxxer is likely to offer the rejoinder, 'But the vaccine is a proven danger, and as much a risk as infection, so I'd rather take my chances that I can avoid both by not being vaccinated.' *If* this were true, who wouldn't decide likewise? But again those inconvenient facts get in the way. As documented earlier, the vaccine is a much lower risk than the disease. But beyond that, vaccination significantly lowers the risk of becoming infected. It is a choice *for* the body's well-being where refusing vaccination is a choice *against* the body's well-being.

Finally, the idea of 'my body, my choice' really only works if the person is resolutely isolated from everyone else. It relies on the fiction that what an individual does to their own body does not have an affect on other bodies. Since we already have touched on matters like infectious transmission, let us simply set that concern to one side.

Instead, let us imagine a situation where the unvaccinated person becomes infected but does not spread the infection to anyone else—a statistical improbability given estimates consistently show a transmission rate of about 2–3 persons by each infected one, but we'll set that side, too. Suppose this infected person becomes ill; this is not guaranteed, but it is likely enough. *Anyone else in*

---

[325] See Santucci, "Unvaccinated," who reports on a 13 state study finding the unvaccinated are 4-5 times more likely to get infected, 10 times more likely to require hospitalization, and 11 times more likely to die.

close relationship, especially a dependent relationship, is disadvantaged even if they avoid infection themselves. First, to protect their own bodies they must keep their distance from the infected loved one. Second, if the infected loved one becomes seriously sick they must suffer alongside them. Third, if the infected loved one is hospitalized they must cope with the fear and anxiety that comes with that, and perhaps be faced with the choice whether to continue care if the person is comatose and intubated. Finally, if the infected loved one dies, they are bereft of the person, the care and love that person provided, and faced with a lifetime of all the consequences of that single choice justified by 'My body, my choice.' Does one's body belong to oneself alone?

An Aside: Pronounced Inconsistency in Reasoning

One ironic matter to note about this third claim is how demonstrative it is of an inconsistency in reasoning that is found so often among anti-vaxxers. The very same people who proclaim this proudly (and loudly) with respect to refusing vaccination often are also the most vocal in opposition to abortion, from which this mantra comes via the feminist pro-choice position.

One of the consequences of relying on emotional appeals and being fixated on a few particular ideas is that it can be easy to miss how different the reasoning appears when extended even a little bit. In other words, a line of reasoning that seems perfectly obvious to an anti-vaxxer about a vaccine might seem silly to them if applied elsewhere, even though the reasoning itself—the 'logic'—is the same.

For example, some anti-vaxxers go so far as to declare that the choice of who lives or dies from COVID-19 should be left up to God and we should therefore shun vaccines. Sometimes this comes across a bit like the argument, 'If God had meant us to fly, he would have given us wings.' After all, technology is responsible for many things in human life, most of which go unchallenged and quite often even unrecognized. Also, the same people who say such things often turn to medical help for injuries and other illnesses.

Being vaccinated or not simply is not magically so important a matter that it warrants a suspension of the normal rules of good thinking. Indeed, whenever stakes are raised the importance of *good* thinking becomes even more critically important.

The Bottom Line

As we indicated much earlier, a minority position is not wrong because it is in the minority. But arguments have to mount beyond, "I disagree." They have to be "I disagree *because* . . ."—and the 'because' has to be more substantive than 'because I've done my research, my authorities are as good as yours, and it's my choice anyway!" The generality and vagueness of most anti-vax arguments aren't helped when the more specific efforts turn out to be riddled with distorted data, misinformation, and false claims. As indicated back in chapter 1, an argument's claims must be validly supported to have merit.

Responses to this Decline

More than once in this book it has been mentioned that not all arguments are equal. The opposing sides in the COVID-19 vaccination debate have each had their proverbial day in court. While each reader will decide for her- or himself which side has fared better, the jury of the American public has largely decided in favor of the pro-vaccination side. However, the strength of the anti-vax movement, which rests in fear-mongering as its tool of persuasion, is such that two outcomes are predictable and actually seen.

On the one hand, some people who were vaccine hesitant but become persuaded to take the vaccine hedge their bets by doing so in secret. The *secretly vaccinated* are motivated by avoidance of anti-vaxxer blowback. According to a Harris poll survey conducted for *USA Today*, approximately 1-in-6 (about 17%) keep their vaccination secret—either from at least some people (about 11%), or from everyone (about 6%). Perhaps even more indicative of the power of the anti-vax movement to arouse anxiety or fear, a full 1-in-4 (25%) of those surveyed reported their vaccination status could cause relationship friction.[326]

On the other hand, hardcore anti-vaxxers still persist with their resistance and double down on their claims. Some anti-vaxxers might be called *denialists*. The term, as we saw earlier, is derived from *denialism*, the use of rhetorical speech designed to preserve the veneer of legitimate debate when, in fact, there is none because the opposing side has carried the day.[327] Thus, despite the overwhelming consensus of experts, accepted by the majority of the general population, denialists continue to maintain their position has unrecognized legitimacy. Why? As suggested elsewhere, this might be because of selfish interests where the anti-vaxxer's livelihood and social prestige are at stake. Or it might be because no one likes to feel duped and the notion that one has been a victim of deceit is so embarrassing, even shameful, that some people will deny it in the face of everything and literally die for their mistaken belief to save themselves a degree of embarrassment.

Is the Anti-Vaccination Movement Treated Fairly?

Anti-vaxxer Mike Yeadon has—correctly—written, "When scientific debate is stifled, people die. Science requires the airing of opinions and debate to allow the evolution of ideas."[328] Unfortunately, he neglects to specify that scientific debate follows established rules no less than high school debate does. Casually setting aside rules, then complaining that one is not being treated fairly, is akin to cheating at cards then griping that being kicked out of the game is 'unfair.' No one wants to play with a card cheat and scientists don't choose to debate with those who ignore the rules of science. Simply making claims in the name of science is not 'debate,' nor is a mere airing of opinions. The latter is certainly

---

[326] Bomey, "Secret Vaxxers."
[327] See Stolle et al. "Fact vs Fallacy," 4484. They contrast the denialist with the 'fair-minded skeptic.'
[328] Yeadon, "The PCR False Positive Pseudo-Epidemic."

happening, but it does not carry with it an entitlement to be viewed as convincing or even worth further attention.

Anti-vaxxers also have a tendency to explain away the rejection of their arguments as the result of Big Pharma and Big Government, in collusion with Big Media, all having actively suppressed their side. Thus one way to handle inconvenient facts is to dismiss their sources as biased and oppressive. So anything said by, for instance, the CDC or FDA can't be trusted because they are Big Government. Any research produced with pharmaceutical funding can't be trusted because it is from Big Pharma. Anything broadcast by the major news outlets can't be trusted because they are just shills for Big Government and Big Pharma.

Unless what is reported can be used by anti-vaxxers to support some claim they make. *Then* it is okay. In other words, anti-vaxxers like to frame the debate in such a fashion that the only voices allowed are their own. They won't trust government *unless* the elected official supports their positions (thank you, GOP!). They won't trust Big Pharma *unless* they support the use of some drug alternative the anti-vaxxer wants (e.g., hydroxychloroquine, remdesivir, or ivermectin). They won't trust Big Media *unless* they grant a measure of credibility to anti-vax claims (thank you, FOX News hosts!).

But let's look more closely at one pet peeve, that the media is unfair. Does media present the vaccination debate fairly? On the one hand, media reports do seem to regularly cite information from public surveys reporting people's attitudes and reasoning, both pro-vaccine and anti-vaccine. This volume has drawn upon such reporting quite often. By citing such data the media can claim to be both factual and neutral.

On the other hand, the media also quite often reports anti-vax views as "myths" or "misrepresentations." Major news outlets tend not to credit anti-vax positions. Is that bias? Should media instead simply report contrasting positions equally and neutrally, letting viewers and readers decide for themselves?

Imagine for a moment obligating journalists to do so every time there is a disagreement. When a story is presented on the Holocaust should it be immediately accompanied by one of equal length by a Holocaust denier? Whenever an image from the international space station shows the earth as a globe should there be an accompanying disclaimer from flat-earthers?

Anti-vaxxers might object that the two cited examples are themselves unfair because the arguments mounted by Holocaust deniers and flat-earthers are so weak. But that is precisely the point: the arguments mounted by anti-vaxxers are so weak and easily refuted they remain convincing only to those who for whatever reason *want to believe them.*

So why are some anti-vaxxers so obstinate? Researchers Stolle and colleagues conclude that denialists are "typically motivated by greed, ideology, eccentricity, and idiosyncrasy."[329] As we have seen, espousing an anti-vax

---

[329] Stolle et al. "Fact vs Fallacy," 4486, citing Diethelm amd McKee, "Denialism."

position can prove lucrative. But while greed may motivate some or all of the well-known leaders of the movement, ideology—being a true believer—is most probably the case for some of their duped followers. I think eccentricity and idiosyncrasy, especially as in a love of being seen as different from others, also motivates some people, who aren't so much convinced by the arguments as they are in love with the attention they get. But mostly I suspect that what is presented as conviction is just a mask covering deep-seated anxiety and fear.

In the real world attitudes are, of course, more mixed than a presentation like this book offers can easily show. For example, research in Israel found that no less than 6 different groups can be distinguished based on attitudes toward vaccination. Although their study addresses specific vaccines in the setting of a different nation, we may not be wrong to adapt their findings to our American setting. There are *judicious acceptors* who favor vaccination as a response to official recommendations; *differentiators* who favor vaccination in some instances but not others; *soft-individualists* who lean toward it being left to personal choice much of the time; *hard-individualists* who favor it being left to personal choice all of the time; *refuters* who reject vaccination; and the *indifferent* who have not formed an opinion, at least not about most vaccines.[330]

But the researchers also make a point we can easily miss: people do not always act in accordance with their basic attitude. That is precisely why some secret vaxxers come into being; they may *say* one thing and *do* another. So in conversing with others we do well to bear in mind they might be espousing what they think they should say—even what they somewhat believe—though in practice they have acted differently. This, by the way, is as true for pro-vaccine advocates as for anti-vaxxers.

The Bottom 'Bottom Line': The Need to Assess Argument Merits

If anti-vaccine arguments are so weak, then why have I spent so much time and effort discussing them? As I said at the beginning of this volume, I did not want to write this book.

But there is a need. *The anti-vax movement costs lives that need not be lost.* As both the historical record and present experience demonstrate, vaccines really are both safe and effective. The unvaccinated are directly and disproportionately hurt by having been deceived by unsubstantiated anti-vaccine claims. Many studies show with numbing regularity the difference in outcomes between the vaccinated and unvaccinated. In fact, the CDC summed up what was known as of early September, 2021, by stating, "Studies so far show that vaccinated people are 8 times less likely to be infected and 25 times less likely to experience hospitalization or death."[331]

Ironically, for a movement that champions capitalism and declares that freedom from vaccines, mask mandates, and social distancing are needed to safeguard the American economy, the actual direct impact of refusing

---

[330] See Velan et al., "Individualism, Acceptance and Differentiation."
[331] CDC, "The Possibility of COVID-19 after Vaccination."

vaccination and ending up in the hospital is staggering. A Kaiser Family Foundation study released in September, 2021, found that just in the preceding three month span from June through August *preventable COVID-19 hospitalizations among unvaccinated adults cost over $5 billion*. Moreover, this burden was not evenly spread across time; each month saw a steeply ascending rate in hospitalizations and attendant costs (32,000 preventable hospitalizations in June; 68,000 in July; 187,000 in August).[332] These numbers not only indicate serious costs at a personal and social level, but an urgent need to address and reduce them.

The danger anti-vaxxers pose is not in being right, but in being wrong in a way that persuades others to their own harm and that of others. Those who remain genuinely undecided, but who lack the skills required to differentiate good arguments from poor ones, may be persuaded by anti-vax rhetoric, especially since it has no reluctance to promote myths and spread lies with apparent missionary zeal. The impassioned insistence of an anti-vaxxer who stirs up anxiety and fear can be much more emotionally persuasive than the calm assurances of scientists who think solid data does not require rhetoric to sell itself.

How real is the danger? Some research suggests that those who are undecided can be persuaded by anti-vax rhetoric in *5–10 minutes*.[333] That speaks more to the power of anxiety and fear desperate for some resolution than it does to American impatience (though that is bad enough itself). However despicable from a moral standpoint, fear-mongering is a very effective persuasive tool—and one generally applied by those who know they can't win an argument based on the merits of their case.

I accept that I may very well be judged to have been unfair in my treatment of anti-vaxxers because what I present shows them in such poor light. But what is one to do when the arguments presented are so weak? Most of what anti-vaxxers offer as 'arguments' are essentially vacuous claims. The level of support is more often than not a few personal anecdotes, a quote from some dubious authority, and manipulation of scientific sources so as to make it appear what they say has a credibility it in fact lacks.

Personally, I would prefer to talk to anti-vaxxers, and about them, purely on the cultural grounds that are really the heart of sincere concern for some anti-vaxxers. I think a much more positive and fruitful discussion could occur divorced from the pseudoscience and misinformation they present as attention-getting window dressing. Because window dressing is all it is. Look carefully at anti-vax arguments and what is found is as quick a pivot as possible to cultural ideology, whether as a selling point or from sincere conviction.

But that is a whole other kettle of fishes as they say. For better or worse, this book has spent a lot of time addressing matters of science, even though I regard the issues of cultural dispute between anti-vaxxers and others just as important.

---

[332] Amin and Cox, "Unvaccinated COVID-19 Hosptalizations."
[333] Stolle et al. "Fact vs Fallacy," 4485.

Chapter 10
Free Advice (Worth the Price!)

Truth be told, like many Americans I am not all that pleased with *any* side arguing over COVID-19 vaccination. It is easy for pro-vaccine advocates to become exasperated with the misinformation, myths, and outright lies-for-profit told by anti-vaxxers. But when that exasperation turns into arrogant, callous, and dismissive treatment of the many anti-vaxxers who are sincerely concerned and anxious, then how can vaccine advocates claim any moral high ground?

At the same time, when the anxiety and fear of anti-vaxxers stops up their ears to even consider the possibility they might be in error on *anything*, then their 'sincere convictions' sound like self-delusion. Simple shouting louder does not make one's position truer.

And those who step aside to wait-and-see, or brandish skepticism like some ultimate truth, might be the worst of all, for in refraining from more direct involvement they contribute nothing positive to resolving the greatest health crisis this generation has seen. In short, there are adequate reasons for all of us to be unhappy with all of us.

Some Free Advice to Pro-Vaccine Advocates

Expert consensus and general public opinion judges the arguments in favor of vaccination to have carried the day against the anti-vaccine movement. Congratulations. So why are you so often defensive in posture? More importantly, why do you continue to rely on persuasive techniques with the undecided that are so much less effective than you desire? It doesn't cut it to simply blame the undecided and refuse to accept any responsibility for poor messaging.

Hey, if what you are doing isn't working, *try something else*. I get that scientists and medical professionals operate within a certain ethos. They resist doing anything they think might diminish the solemn, calm, assured, authoritative expertise they like to project. But some of the most effective and successful scientists of all time have been passionate articulators, like Albert Einstein.

It is alright to speak truth *emotionally*. Most people are not persuaded by 'the facts, ma'am, just the facts'—as the detective of the old television series *Dragnet* used to demand. Sacrifice of facts—of reason and evidence—is not required when one also makes an appeal to emotions. In fact, it can underscore the compelling power of the science being presented.

The fact of the pandemic is that it is *scary*. Science does not need to shield us from that. In fact, though, what pro-vaccine advocates have tended to do is send a mixed message, speaking out of both sides of the mouth, first telling us of the dire consequences of not being vaccinated then rushing to assure us that individual risk of dire consequences is very low. Both claims may be true, but can't they be presented in a way that helps people understand how both can be true?

Pro-vaccine arguments are heavily weighted to address scientific concerns and in a scientific manner. This is consistent with what we saw in an earlier chapter as to what pro-vaccine advocates regard as the heart of the matter. But is it enough? That approach is highly inaccessible to most people. In fact, it is easy for anti-vaxxers to capitalize on that fact by presenting the illusion of science because *most people cannot distinguish science fact from science fiction.*

There is a place for highly technical literature filled with precise scientific terminology. It is absolutely vital literature. But it is not easily understood and it is okay for those with the scientific knowledge and ability to do so to 'dumb it down' for the rest of us. No one is asking for unsubstantiated generalizations or sweeping promises, but much science can be made more accessible and without all the hemming, hawing, and hedging that those in-the-know will already presume (if they are bothering to read popular presentations like I'm talking about).

As we saw earlier, substantial research suggests that once an individual has been persuaded to take an anti-vaccine position it is difficult to reverse course. Perhaps, then, the most substantial criticism that can be made against the pro-vaccine movement is its relative failure to do better in promoting vaccines *before* a crisis emerges and anti-vax rhetoric ramps up.

What is needed to protect and further vaccine efforts is a plan to inoculate people against anti-vaccine conspiracy theories *before* they are encountered. An old saying proclaims, 'an ounce of prevention is worth a pound of cure.' In the case of protecting people against misinformation, myths, and groundless conspiracy theories, an active program of science-based, accessible, and easily understand materials is clearly desirable.[334]

But so, too, is addressing those things anti-vaxxers regard as more important than science. *Pro-vaccine arguments need to extend to address anti-vax culture*—especially about political rights like individual liberty and religious concerns about science arrogating to itself an authority and knowledge it does not legitimately possess. Borrow a page from the anti-vaxxers (hopefully without the cynicism implicit in some of it) and talk more about how science supports values Americans hold dear.

Unfortunately, scientists are generally neither well-disposed nor adequately equipped to make such arguments. Instead, the tendency has been to dismiss cultural concerns as either insignificant or completely beside the point. Such tunnel vision undermines the power of the science of the pro-vaccine position.

Fortunately, there is hope. The persistent success of anti-vaxxers in sowing unwarranted doubt is prompting more and more pro-vaccine people to reexamine their approach and to try to modify it to become more effective. It is about time.

---

[334] See Jolly and Douglas, "Prevention Is Better than Cure."

Some Free Advice to Anti-Vaxxers

None of the criticism of the anti-vaccine movement's arguments excludes the possibility anti-vaxxers might be right in at least some respects. But we can still say that their *arguments* are weak, that the arguments' weaknesses undermine anti-vax credibility, and that a lack of credibility makes it hard for their position to be taken as seriously as they would like. So the bottom-line advice is this: learn to distinguish *credible* claims from *incredible* ones—and stay to the former.

Two possibilities exist for anti-vaxxers: either find better arguments, or reconsider whether those arguments are as right as they want them to be. Frankly, after much concerted searching over a significant period of time, I found no arguments by anti-vaxxers purporting to be 'science' credible under scrutiny. They consistently misused sources they alluded to, whether intentionally or simply from a lack of understanding what they were reading and discussing is not for me to say. But *half-truths are full lies*.

Jonathan Berman, in his book *Anti-Vaxxers*, gets to the heart of the problem when discussing pro-vaccine and anti-vaccine positions together.

> Rarely do issues have 'two sides' of equal scientific merit that deserve equal representation. Credentials alone mean little. Those representing themselves as scientists or physicians may well have doctorates, but they may also be speaking well outside their areas of expertise. The story of the little guy going up against an evil corporation may make for a compelling (and well-worn) narrative, but often the little guy is working with bad science.[335]

Biochemist Lucas Stolle and colleagues observe that anti-vax arguments rely too often on stories (anecdotal evidence), leaving them prone to the logical error of presuming that because *this* happened *that* must be the reason. They also frequently present false correlations but present them as facts.[336] So a good place to start for making better arguments is to *stop doing what isn't credible*.

Beyond that, Stolle and colleagues also point to apparent biases among anti-vaxxers that can compromise sound reasoning. These include a willingness to believe something despite a lack of evidence; favoring arguments merely because they support what one already believes; accepting false or illusory correlations; prizing some experiences or events at the expense of others that might be more valid or relevant, and resolving conflicts between pieces of evidence by choosing whichever one reduces the conflict regardless of any other merits.[337]

Of course, changing direction is easier said than done for many things, and where an emotional investment has been made some people will choose to 'double-down' in their endorsement rather than risk the horror of admitting having made a mistake. In fact, physician and professor Kenneth Carmago, Jr.

---

[335] Berman, *Anti-Vaxxers*, 83.
[336] Stolle et al. "Fact vs Fallacy," 4483.
[337] Stolle et al. "Fact vs Fallacy," 4484. Formally, these are, in order: bias of omission, confirmation bias, correlation errors, the availability heuristic, and resolving cognitive dissonance.

calls attention to a 'backfire effect' in which someone who embraced a mistaken notion simply reacts to counter evidence by distorting things such as to use the new information to reinforce the preexisting belief.[338] Put another way, like the philosopher Hegel, who allegedly declared, 'If the facts disagree with my theory, too bad for the facts!', some people are quite willing to dispense with knowledge and truth if it serves their ego and quells personal discomfort.

Unfortunately, as is often the case with small groups with strident voices, there is among anti-vaxxers a pronounced willingness to let a few authoritarian figures set out the case and then followers merely repeat it as loud as possible—a variant of the philosophy often found among spouses that 'If I only yell louder I'll be heard and convince my partner.' How well does that work?

But let us suppose one is willing to do some further research. The word literally means "look ('search') again ('re-')." Even that can present problems. First, one already has searched and found plenty of repetition of what they thought was true but are now reconsidering. Yet looking for new information proves not so easy. As Stolle and colleagues note, search engines are built based on what one has already done, so search results tend to favor reproducing the same and similar results to what one found before. This phenomenon itself can produce an illusory situation reinforcing confirmation bias.[339]

Moreover, studies have shown that many people have great difficulty distinguishing accurate information from inaccurate. As we have seen, 'figures don't lie, but liars figure.' Numbers and quotes from experts divorced from context can be made to spread inaccurate messages. The difficulty many folk can have is shown in one study of 34 students, all undecided about vaccination, who were directed in internet searching to 40 different websites, all of which presented supposed 'medical facts' about vaccination. Only 18 of the 40 presented accurate information overall but 20 (59%) of the students concluded *all* of the websites were, on the whole, accurate. Fortunately, subsequent exposure to accurate, verified information proved helpful in the students then being able to better separate accurate from inaccurate information.[340]

Regrettably, too many Americans are scientifically illiterate. This is a situation anti-vaxxers exploit repeatedly to great effect. Even those who are college educated and presumably taught critical thinking may struggle to differentiate what is scientifically credible. Anti-vaxxers are happy to 'interpret' scientific articles for them, relying on their ignorance in order to say whatever they want, secure in the knowledge they won't be called out by those who already believe them because they want to believe them.

Pro-vaccine advocates simply *must* do better at making science accessible. They must help people discern sense from nonsense. It is my hope this book assists in promoting that kind of progress.

---

[338] Camargo Jr., "Here We Go Again," 3, citing Cook and Lewandowsky, *The Debunking Handbook*.
[339] Stolle et al. "Fact vs Fallacy," 4485.
[340] Kortum, Edwards, and Richards-Khortum, "The Impact of Inaccurate Internet Health Information."

Some Free Advice to Those 'Who Don't Take Sides'

A final piece of advice has to be directed to the skeptics and the undecided—the 'vaccine hesitant.' The present pandemic has been likened to a war with the COVID-19 virus. I can understand the undecided person's wish or desire not to get involved with the civil war between anti-vaxxers and pro-vaccine advocates.

But frankly, that's the lesser war. The real and ultimately most significant battle is against the virus. One can hardly be neutral in that fight. In fact, as has been pointed out earlier, in this larger war to not choose a side between vaccination and natural infection is, in truth, to have chosen natural infection.

I don't mean that vaccine refusal is a guarantee of becoming infected with COVID-19. But it does certainly increase the likelihood compared to choosing vaccination. Voting with one's feet is the actual way any destination is chosen. One can spend as much time as is desired poring over maps to figure out the best route to get somewhere, but where one's feet actually are at is where the real destination remains.

So know that you are choosing a side already. The question remains: Is this a choice you are happy risking your health and life (and that of loved ones) on?

Final Free Advice—from Someone Else

I am going to let someone else have the final word.

Daryl Rise, an Idahoan whose mother had a severe case of COVID-19 that put her in a coma, and whose sister died of the disease after urging others not to get vaccinated and refusing vaccination herself, said in an interview, "We're hearing from all these doctors and professionals who have all this education and they're basically begging us to get the vaccination. The people that are telling us not to, they're not as educated as these doctors, and they're following social media. It doesn't matter if we're a donkey or an elephant. It is a personal choice, but the numbers don't lie."[341]

---

[341] Matthew, "Idaho Nurse Who Refused COVID Vaccine Dies."

Personal Notes

Appendix

Joint Statement in Support of COVID-19 Vaccine Mandates for All Workers in Health and Long-Term Care
(July 26, 2021)

<u>Signees</u> (58 professional medical groups)

Academy of Managed Care Pharmacy (AMCP)
American Academy of Ambulatory Care Nursing (AAACN)
American Academy of Child and Adolescent Psychiatry (AACAP)
American Academy of Family Physicians (AAFP)
American Academy of Nursing (AAN)
American Academy of Ophthalmology (AAO)
American Academy of PAs (AAPA)
American Academy of Pediatrics (AAP)American Association of Allergy, Asthma & Immunology (AAAAI)
American Association of Clinical Endocrinology (AACE)
American Association of Colleges of Pharmacy (AACP)
American Association of Neuroscience Nurses (AANN)
American College of Clinical Pharmacy (ACCP)
American College of Physicians (ACP)
American College of Preventive Medicine (ACPM)
American College of Surgeons (ACS)
American Epilepsy Society (AES)
American Medical Association (AMA)
American Nurses Association (ANA)
American Pharmacists Association (APhA)
American Psychiatric Association (APA)
American Public Health Association (APHA)
American Society for Clinical Pathology (ASCP)
American Society for Radiation Oncology (ASTRO)
American Society of Health-System Pharmacists (ASHP)
American Society of Hematology (ASH)
American Society of Nephrology (ASN)
American Thoracic Society (ATS)
Association for Clinical Oncology (ASCO)
Association for Professionals in Infection Control and Epidemiology (APIC)
Association of Academic Health Centers (AAHC)
Association of American Medical Colleges (AAMC)
Association of Rehabilitation Nurses (ARN)
Council of Medical Specialty Societies (CMSS)
HIV Medicine Association
Infectious Diseases Society of America (IDSA)
LeadingAge
National Association of Indian Nurses of America (NAINA)
National Association of Pediatric Nurse Practitioners (NAPNAP)
National Council of State Boards of Nursing (NCSBN)
National Hispanic Medical Association (NHMA)
National League for Nursing (NLN)
National Medical Association (NMA)
National Pharmaceutical Association (NPhA)
Nurses Who Vaccinate (NWV)

Organization for Associate Degree Nursing (OADN)
Pediatric Infectious Diseases Society (PIDS)
Philippine Nurses Association of America, Inc (PNAA)
Society of Gynecologic Oncology (SGO)
Society for Healthcare Epidemiology of America (SHEA)
Society of Hospital Medicine (SHM)
Society of Infectious Diseases Pharmacists (SIDP)
Society of Interventional Radiology (SIR)
Texas Nurses Association (TNA)
The John A. Hartford Foundation
Transcultural Nursing Society (TCNS)
Virgin Islands State Nurses Association (VISNA)
Wound, Ostomy, and Continence Nurses Society (WOCN)

Bibliography

AAPS (Association of American Physicians and Surgeons). "Statement in Support of the Right of All, Including Medical Workers, to Decline Medical Intervention." Accessed online at https://aapsonline.org/statement-in-support-of-the-right-of-all-including-medical-workers-to-decline-medical-intervention/.

ABC News. "COVID Death Toll in US Eclipses 1918 Influenza Pandemic Estimates." (Sept. 20, 2021). Accessed online at https://www.msn.com/en-us/health/medical/covid-death-toll-in-us-eclipses-1918-influenza-pandemic-estimates/ar-AAOE3va?ocid=winp1taskbar.

Aleccia, JoNel. "A Pill to Treat COVID-19: 'We're Talking about a Return to, Maybe, a Normal Life." CNN (Sept. 27, 2021). Accessed online at https://www.msn.com/en-us/health/medical/ a-pill-to-treat-covid-19-we-re-talking-about-a-return-to-maybe-normal-life/ar-AAOSro2?ocid=msedgdhp&pc=U531.

Amanat, Fatima, and Florian Krammer. "SARS-CoV-2 Vaccines: Status Report." *Immunity* 52 (Apr. 14, 2020) 583–89.

AMA (American Medical Association). "AMA, APhP, ASHP Statement on Ending Use of Ivermectin to Treat COVID-19." AMA (Sept. 1, 2021). Accessed online at https://www.ama-assn.org/press-center/press-releases/ ama-apha-ashp-statement-ending-use-ivermectin-treat-covid-19.

American Public Health Association (APHA). "Joint Statement in Support of COVID-19 Vaccine Mandates for All Workers in Health and Long-term Care." Accessed online at https://www.apha.org/News-and-Media/News-Releases/ APHA-News-Releases/2021/COVID-19-vaccine-mandates

Amin, Krutika, and Cynthia Cox. "Unvaccinated COVID-19 Hospitalizations Cost Billions of Dollars." Peterson-KFF Health System Tracker (Sept. 14, 2021). Accessed online at https://www.healthsystemtracker.org/brief/unvaccinated-covid-patients-cost-the-u-s- health-system-billions-of-dollars/.

Antonelli, Michela, et al. "Risk Factors and Disease Profile of Post-Vaccination SARS-CoV-2 Infection in UK Users of the COVID Symptom Study app: A Prospective, Community-based, Nested, Case-Control Study." *Lancet Infectious Diseases* (2021). Published online September 1, 2021. Accessed online at https://www.thelancet.com/action/showPdf?pii= S1473-3099%2821%2900460-6.

AP-NORC. "Majorities Support Vaccine Mandates for Some Activities Amidst Delta Surge." Poll (Aug. 20, 2021). Accessed online 8/29/2021 at https://apnorc.org/projects/majorities-support-vaccine-mandates-for-some-activities-amidst-delta-surge/.

Arivarasan, Vishnu Kirthi, Karthik Loganathan, and Pushpamalar Janarthanan, eds. *Nanotechnology in Medicine*. Nanotechnology in the Life Sciences. Switzerland: Springer Nature Switzerland, 2021.

Atlas, Scott W. "The Data Is in—Stop the Panic and End the Total Isolation." *The Hill* (Apr. 22, 2020). Accessed online at https://thehill.com/opinion/healthcare/ 494034-the-data-are-in-stop-the-panic-and-end-the-total-isolation.

Baraniuk, Chris. "How Long Does Covid-19 Immunity Last?" *BMJ* 373 (2021) No. 1605. Accessed online at https://www.bmj.com/content/bmj/373/bmj.n1605.full.pdf

Baxby, Derrick. "Edward Jenner's Role in the Introduction of Smallpox Vaccine." In *History of Vaccine Development*, edited by Stanley A. Plotkin, 13–19. New York: Springer, 2011.

Beigel, John H., et al. "Remdesivir for the Treatment of Covid-19—Final Report." *New England Journal of Medicine* 383 (2020) 1813–26.

Beleche, Trinidad, et al. "COVID-19 Vaccine Hesitancy: Demographic Factors, Geographic Patterns, and Changes Over Time." ASPE (Assistant Secretary for Planning and Evaluation) Issue Brief (May, 2021). Accessed online at https://aspe.hhs.gov/sites/default/files/private/pdf/265341/aspe-ib-vaccine-hesitancy.pdf.

Bellware, Kim, and Drea Cornejo, "How Pastors and Health Experts Are Struggling to Close the Vaccine Gap among White Evangelicals." *The Washington Post* (June 18, 2021). Accessed online at https://www.washingtonpost.com/religion/2021/06/18/white-evangelicals-vaccine/.

Berg, Sara. "How a Decade of Coronavirus Research Paved Way for COVID-19 Vaccines." AMA (Feb. 26, 2021). Accessed online at https://www.ama-assn.org/delivering-care/public-health/ how-decade-coronavirus-research-paved-way-covid-19-vaccines.

Berman, Jonathan M. *Anti-Vaxxers: How to Challenge a Misinformed Movement*. Cambridge: MIT Press, 2020.

Bernigaud, C., et al. "Oral Ivermectin for a Scabies Outbreak in a Long-term Care Facility: Potential Value in Preventing COVID-19 and Associated Mortality." *British Journal of Dermatology* 184 (2021) 1207–09. Accessed online at https://onlinelibrary.wiley.com/doi/epdf/10.1111/bjd.19821.

BioLogos. "Should Christians Get Vaccinated?" (Updated May 7, 2021). Accessed online at https://biologos.org/common-questions/should-christians-get-vaccinated.

Blake, Aaron. "How Badly Unvaccinated Republicans Are Misinformed, in One Stat." *The Washington Post* (Sept. 27, 2021). Accessed online at https://www.msn.com/en-us/news/politics/how-badly-unvaccinated-republicans-are-misinformed-in-one-stat/ar-AAOSkBZ?ocid=msedgntp.

———. "How the Right's Ivermectin Conspiracy Theories Led to People Buying Horse Dewormer." *The Washington Post* (Aug. 24, 2021). Accessed online at https://www. washingtonpost.com/politics/2021/08/24/how-rights-ivermectin-conspiracy-theories-led-people-buying-horse-dewormer/.

———. "Sherri Tenpenny's Bizarre Anti-Vaccine Testimony Was a Long Time Coming." *The Washington Post* (June 9, 2021). Accessed online at https://www.washingtonpost.com/politics/2021/06/09/sherri-tenpennys-bizarre-anti-vaccine-testimony-was-long-time-coming/.

Blume, Stuart. *Immunization: How Vaccines Became Controversial*. London: Reaktion Books, 2017.

Bolich, G. G. *Two Masters: Evangelicals and the GOP*. Spokane: EVS Press, 2020.

———. *Knowledge: An Illustrated History*. Spokane: EVS Press, 2019.

———. *Knowledge and Belief: Their Natures and Relationship in Ancient Greek Thought*. EVS Press, 2019.

———. *Understanding Arguments: The SIMPLE Method*. EVS Press, 2019.

Bomey, Nathan. "Secret Vaxxers: These Americans Are Getting COVID Vaccinations But Not Telling Anyone." *USA Today* (Sept. 2, 2021). Accessed online at https://www.msn.com/en-us/news/us/secret-vaxxers-these-americans-are-getting-covid-vaccinations-but-not-telling-anyone/ar-AAO1SiX?ocid=msedgdhp&pc=U531.

Brown, Catherine M., et al. "Outbreak of SARS-CoV-2 Infections, Including COVID-19 Vaccine Breakthrough Infections, Associated with Large Public Gatherings—Barnstable County, Massachusetts, July 2021." CDC *Morbidity and Mortality Weekly Report* 70/31 (Aug. 6, 2021) No. 31. Accessed online at https://www.cdc.gov/mmwr/volumes/70/wr/pdfs mm7031e2-H.pdf.

Brownstein, Barry. "COVID-19 Lockdowns May Destroy Our Immune Systems." *Chronicles Magazine* [Charlemagne Institute, Intellectual Takeout website] (Apr. 29, 2020). Accessed online at https://www.intellectualtakeout.org/covid-19-lockdowns-may-destroy-our-immune-systems/?fbclid=IwAR3DTEJabeH_b2ep06T1ue8zQD9X-7aXAriKJtqpM-rAnKYLvRMGctMX f8E.

Brumfiel, "Anti-Vaccine Activists Use a Federal Database to Spread Fear about COVID Vaccines." NPR (National Public Radio) (June 14, 2021). Accessed online at https://www.npr.org/ sections/health-shots/2021/06/14/1004757554/anti-vaccine-activists-use-a-federal-database-to-spread-fear-about-covid-vaccine.

Bryant, Andrew, et al. "Ivermectin for Prevention and Treatment of COVID-19 Infection: A Systematic Review, Meta-analysis, and Trial Sequential Analysis to Inform Clinical Guidelines." *American Journal of Therapeutics* 28 (2021) e434–60. Accessed online at Ivermectin_for_Prevention_and_Treatment_of.7.pdf.

Buonaguro, Franco M., Paulo A. Ascierto, and Gene D. Morse, "Covid-19: Time for a Paradigm Change [Editorial]." *Reviews in Medical Virology* (2020) e2134. Published online July 3, 2020. Accessed online at https://www.ncbi.nlm.nih.gov/pmc/articles/PMC7361272/pdf/RMV-9999-e2134.pdf.

Burki, Talha. "The Online Anti-Vaccine Movement in the Age of COVID-19." *The Lancet: Digital Health* 2 (2020) e504–05.

Butler, Kiera. "Anti-Vaxxers Have a Dangerous Theory Called 'Natural Immunity' Now It's Going Mainstream." *Mother Jones* (May 12, 2020). Accessed online at https://www.motherjones .com/politics/2020/05/anti-vaxxers-have-a-dangerous-theory-called-natural-immunity-now-its-going-mainstream/

Butler-Laporte, Guillaume, et al. "Vitamin D and COVID-19 Susceptibility and Severity in the COVID-19 Host Genetics Initiative: A Mendelian Randomization Study." *PLOS Medicine* (June 1, 2021). Accessed online at https://journals.plos.org/plosmedicine/article?id=10.1371/journal.pmed.1003605.

Caldwell, Alison. "Study Suggests High Vitamin D Levels May Protect against COVID-19, Especially for Black People." UchicagoMedicine (Mar. 19, 2021). Accessed online at https://www.uchicagomedicine.org/forefront/coronavirus-disease-covid-19/vitamin-d-covid-study.

Callaghan, Timothy, et al. "Correlates and Disparities of Intention to Vaccinate against COVID-19." *Social Science and Medicine* 272 (2021) 113638. Accessed online at https://www.ncbi.nlm.nih.gov/pmc/articles/PMC7834845/pdf/main.pdf.

Cao, Yuhong, and George F. Gao. "mRNA Vaccines: A Matter of Delivery." *Eclinical Medicine* 32 (2021) 100746. Accessed online at https://www.thelancet.com/action/showPdf?pii=S2589-5370%2821%2900026-2.

Cappelan, Alexander W., et al. "Choice and Personal Responsibility: What Is a Morally Relevant Choice?" *Review of Economics and Statistics* (2020) 1-35. Accessed online at https://direct.mit.edu/rest/article/doi/10.1162/rest_a_01010/97740/Choice-and-Personal-Responsibility-What-Is-a.

Carmago Jr., Kenneth Rochel de. "Here We Go Again: The Re-emergence of Anti-Vaccine Activism on the Internet." *CSP (Cadernos de Saude Publica/Reports on Public Health)* 36/Sppl. 2 (2020) e00037620.

CCDH (Center for Countering Digital Hate). "The Disinformation Dozen." CCDH (2021). Accessed online at f4d9b9_b7cedc0553604720b7137f8663366ee5.pdf.

———. *Pandemic Profiteers: The Business of Anti-Vax.* Accessed online at https://252f2edd-1c8b-49f5-9bb2-cb57bb47e4ba.filesusr.com/ugd/f4d9b9_13cbbbef105e459285ff21e94ec34157.pdf.

CDC (Centers for Disease Control and Prevention). "COVID-19 Vaccine Breakthrough Case Investigating and Reporting." CDC: Vaccines and Immunizations (Aug. 25, 2021). Accessed online at https://www.cdc.gov/vaccines/covid-19/health-departments/breakthrough-cases.html.

———. "Frequently Asked Questions about COVID-19 Vaccination." (Updated Sept. 3, 2021). Accessed online at https://www.cdc.gov/coronavirus/2019-ncov/vaccines/faq.html?s_cid=10492:what%27s%20in%20the%20covid%2019%20vaccine:sem.ga:p:RG:GM:gen:PTN:FY21.

———. "New CDC Study: Vaccination Offers Higher Protection than Previous COVID-19 Infection." (Aug. 6, 2021). Accessed online at https://www.cdc.gov/media/releases/2021/s0806-vaccination-protection.html.

———. "The Possibility of COVID-19 after Vaccination: Breakthrough Infections." CDC/COVID-19 (Updated Sept. 7, 2021). Accessed online at https://www.cdc.gov/coronavirus/2019-ncov/vaccines/effectiveness/why-measure-effectiveness/breakthrough-cases.html/.

———. "Selected Adverse Events Reported after COVID-19 Vaccination." (Updated Sept. 2, 2021). Accessed online at https://www.cdc.gov/coronavirus/2019-ncov/vaccines/safety/adverse-events.html.

———. "Symptoms of COVID-19." (Updated Feb. 22, 2021). Accessed online at https://www.cdc.gov/coronavirus/2019-ncov/symptoms-testing/symptoms.html.

———. "Thimerosal and Vaccines." Accessed online at https://www.cdc.gov/vaccinesafety/concerns/thimerosal/index.html.

———. "Understanding mRNA COVID-19 Vaccines." (Updated Mar. 4, 2021). Accessed online at https://www.cdc.gov/coronavirus/2019-ncov/vaccines/different-vaccines/mrna.html.

———. "Understanding Viral Vector COVID-19 Vaccines." (Updated Apr. 13, 2021). Accessed online at https://www.cdc.gov/coronavirus/2019ncov/vaccines/different-vaccines/viralvector.html.

CDC/FDA. "Allergic Reactions Including Anaphylaxis after Receipt of the First Dose of Modern COVID-19 Vaccine—United States, December 21, 2020–January 10, 2021." CDC *Morbidity and Mortality Weekly Report* 70/4 (Jan. 29, 2021) 125–29. Accessed online at https://www.ncbi.nlm.nih.gov/pmc/articles/PMC7842812/pdf/mm7004e1.pdf.

Chen, Frederick, and Flavio Toxvaerd. "The Economics of Vaccination." *Journal of Theoretical Biology* 363 (2014) 105–17.

Children's Hospital of Philadelphia (CHOP). "Vaccine Ingredients—Thimerosal." Accessed online at https://www.chop.edu/centers-programs/vaccine-education-center/vaccine-ingredients/thimerosal

Cho, Alice, et al. "Antibody Evolution after SARS-CoV-2 mRNA Vaccination." bioRxiv (2021). Accessed online at https://www.biorxiv.org/content/10.1101/2021.07.29.454333v2.full.

Christians and the Vaccine. "About Us." Accessed online at https://www.christiansandthevaccine.com/about.

———. "FAQs about Christians and the Vaccine." Accessed online at https://www.christiansandthevaccine.com/faqs.

Commisso, Danielle. "Vaccine Hesitancy Exists in the U.S., But Majority of U.S. Adults Believe in Vaccinating Children." CivicScience. (Feb. 6, 2019). Accessed online at https://civicscience.com/vaccine-hesitancy-exists-in-the-u-s-but-majority-of-u-s-adults-believe-in-vaccinating-children/.

Cook, J., and S. Lewandowsky. *The Debunking Hand-book*. St. Lucia: University of Queensland, 2011.

Crary, David. "Vaccine Skepticism Runs Deep among White Evangelicals in US." NBC15News (July 29, 2021). Accessed online at https://mynbc15.com/news/local/vaccine-skepticism-runs-deep-among-white-evangelicals-in-us-07-29-2021.

Cronin, "'It's Not Looking in Our Favor': man Who Organized Anti-Mask Freedom Rally Now on a Ventilator after Catching COVID." *Daily Dot* (Aug. 28, 2021). Accessed online 8/28/2021 at https://www.msn.com/en-us/news/us/it-s-not-looking-in-our-favor-man-who-organized-anti-mask-freedom-rally-now-on-a-ventilator-after-catching-covid/ar-AANOPmv?ocid=msedgntp.

Cupuk, Jasmine, et al. "The SARS-CoV-2 Nucleocapsid Protein Is Dynamic, Disordered, and Phase Separates with RNA." *Nature Communications* 12 (2021) 1936. Accessed online at https://www.nature.com/articles/s41467-021-21953-3.pdf.

Dan, Jennifer M., et al. "Immunological Memory to SARS-CoV-2 Assessed for Up to 8 Months after Infection." *Science* 371 (Feb. 5, 2021) eabf 4063. Accessed online at https://www.science.org/doi/epdf/10.1126/science.abf4063.

Daniels, Deborah. "Vaccines Are Not Just a Matter of Personal Choice." *Indianapolis Business Journal* (Aug. 13, 2021). Accessed online at https://www.ibj.com/articles/deborah-daniels-vaccines-are-not-just-a-matter-of-personal-choice.

Dastagir, Alia E. "Facts Alone Don't Sway Anti-Vaxxers. So What Does?" *USA Today* (Apr. 8, 2019). Accessed online at https://www.usatoday.com/story/news/investigations/2019/03/08/vaccine-anti-vax-anti-vaxxer-what-change-their-mind-vaccine-hesitancy/3100216002/.

Davidson, Tish. *Vaccines. History, Science, and Issues*. Santa Barbara: Greenwood, 2017.

Davies, P., S. Chapman, and J. Leask. "Antivaccination Activists on the World Wide Web." *Archives of Disease in Childhood* 87/1 (2002) 22–25.

Dawson, Bethany. "Anti-vaxxers Are Gargling Iodine in the Latest Ill-Advised Attempt at DIY Anti-COVID Care, Say Reports." Business Insider (Sept. 26, 2021). Accessed online at https://www.businessinsider.com/anti-vaxxers-are-gargling-iodine-try-and-stop-covid-19-2021-9.

Dean, Claudia, Kim Parker, and John Gramlich. "A Year of U.S. Public Opinion on the Coronavirus Pandemic." Pwe Research Center (Mar. 5, 2021). Accessed online at https://www. pewresearch.org/2021/03/05/a-year-of-u-s-public-opinion-on-the-coronavirus-pandemic/.

Delaney, Patrick. "EXCLUSIVE: Former Pfizer VP: 'Your Government Is Lying to You in a Way that Could Lead to Death." LifeSiteNews (Apr. 7, 2021). Accessed online at https://www.lifesitenews.com/news/exclusive-former-pfizer-vp-your-government-is-lying-to-you-in-a-way-that-could-lead-to-your-death/.

Diethelm, P., and M. McKee. "Denialism: What Is it and How Should Scientists Respond?" *European Journal of Public Health* 19/1 (2009) 2–4.

Dobson, James. "The Unprecedented Production of the COVID-19 Vaccines Due to Operation Warp Speed and the Use of Aborted Fetal Cell Lines." Statement from The Dr. James Dobson Family Institute (Apr. 5, 2021). Accessed online at https://www.drjamesdobson.org/news/unprecedented-production-covid-19-vaccines-due-operation-warp-speed-and-use-aborted-fetal-cell.

Dopico, Xaquin Castro, et al. "Immunity to SARS-CoV-2 Induced by Infection or Vaccination." *Journal of Internal Medicine* (2021) 0, 1–19. Accessed online at https://onlinelibrary.wiley .com/doi/epdf/10.1111/joim.13372

Du, Zhanwei, et al. "Serial Interval of COVID-19 among Publicly Reported Confirmed Cases." *Emerging Infectious Diseases* 26/6 (June, 2020) 1341–43.

Ducharme, Jamie. "The Delta Variant Is Pushing Pediatric Hospitals to the Brink." *Time* (Sept. 13/Sept. 20, 2021) 9–10.

Durbach, Nadja. *Bodily Matters: The Anti-Vaccine Movement in England, 1853–1907.* Durham: Duke Univ. Press, 2005.

Elgazzar, Ahmed, et al., "Efficacy and Safety of Ivermectin for Treatment and Prophylaxis of COVID-19 Pandemic." Research Square. (2020) Accessed online at https://assets. researchsquare.com/files/rs-100956/v3/52188c0f-1239-455b-9d58-82041615f02c.pdf?c=1629143760.

Ellenburg, Susan, et al. "The Long View on COVID-19 Vaccine Safety and Efficacy." PennToday (July 13, 2021). Accessed online at https://penntoday.upenn.edu/news/long-view-covid-vaccine-safety-and-efficacy.

Ellis, Phoebe, et al. "Decoding Covid-19 with the SARS-CoV-2 Genome." *Current Genetics Medicine Report* 9 (2021) 1–12.

Faust, Jeremy Samuel, and Carlos del Rio. "Assessment of Deaths from COVID-19 and from Seasonal Influenza [Viewpoint]." *JAMA Internal Medicine* 180/8 (2020) 1045–46.

Fauzia, Miriam. "Fact Check: The Vaccine for COVID-19 Has Been Nearly 20 Years in the Making." *USA Today* (Jan. 21, 2021). Accessed online at https://www.usatoday.com/story/news/factcheck/2021/01/21/fact-check-covid-19-vaccine-nearly-20-years-making/3873247001/.

FDA (Food and Drug Administration). "FAQ: COVID-19 and Ivermectin Intended for Animals." (Apr. 26, 2021). Accessed online at https://www.fda.gov/animal-veterinary/product-safety-information/faq-covid-19-and-ivermectin-intended-animals.

———. "FDA Cautions against Use of Hydroxychloroquine or Chloroquine for COVID-19 Outside of the Hospital Setting or a Clinical Trial Due to Risk of Heart Rhythm Problems." (Update of July 1, 2020). Accessed online at https://www.fda.gov/drugs/drug-safety-and-availability/fda-cautions-against-use-hydroxychloroquine-or-chloroquine-covid-19-outside-hospital-setting-or.

———. "Vaccine Information Fact Sheet for Recipients and Caregivers about Comirnaty (COVID-19 Vaccine, mRNA) and Pfizer-Biontech COVID-19 Vaccine to Prevent Coronavirus Disease 2019(COVID-19). Accessed online at https://www.fda.gov/media/144414/ download.

Feldman, Justin. "All the Ways That '1 in 5,000 per Day' Breakthrough Infection Stat Is Nonsense." *Slate* (Sept. 25, 2021). Accessed online at https://www.msn.com/en-us/health/medical/all-the-ways-that-1-in-5-000-per-day-breakthrough-infection-stat-is-nonsense/ar-AAONRV0?ocid=msedgdhp&pc=U531.

Fontanet, Arnaud, and Simon Caucheme. "COVID-19 Herd Immunity: Where Are We?" *Nature Reviews* 20 (Oct., 2020) 583–84. Accessed online at https://www.nature.com/articles/ s41577-020-00451-5.pdf.

Frenkel, Sheera. "The Most Influential Spreader of Coronavirus Misinformation Online." *New York Times* (July 24, 2021). Accessed online 8/29/31 at https://www.nytimes.com/2021/ 07/24/technology/joseph-mercola-coronavirus-misinformation-online.html.

Federation of State Medical Boards (FMSB), "FMSB: Spreading COVID-19 Vaccine Misinformation May Put Medical License at Risk." FSMB (July 29, 2021). Accessed online at https://www.fsmb.org/advocacy/news-releases/fsmb-spreading-covid-19-vaccine-misinformation-may-put-medical-license-at-risk/.

Fung, Katherine. "7 Myths about the COVID Vaccines Debunked." *Newsweek* (online publication Aug. 27, 2021), accessed online at https://www.msn.com/en-us/health /medical/7-myths-about-the-covid-vaccines-debunked/ arAANOsJY?ocid=msedgdhp&pc =U531.

Funk, Cary, and Becka A. Alper. "Religion and Science." Pew Research Center Oct. 22, 2015). Accessed online at https://www.pewresearch.org/internet/wp-content/uploads/ sites/9/2015/10/PI_2015-10-22_religion-and-science_FINAL .pdf.

Funk, Cary, and John Gramlich, "10 Facts about Americans and Coronavirus Vaccines." Pew Research Center (Mar. 23, 2021). Accessed online at https://www.pewresearch.org/fact-tank/2021/03/23/10-facts-about-americans-and-coronavirus-vaccines/.

Funk, Cary, and Alec Tyson. "Growing Share of Americans Say They Plan to Get a COVID-19 Vaccine—Or Already Have." Pew Research Center (Mar. 5, 2021). Accessed online at file:///C:/Users/Owner/AppData/Local/Temp/PS_ 2021.03.05_covid-19-vaccines_REPORT.pdf.

———. "Intent to Get a COVID-19 Vaccine Rises to 60% as Confidence in Research and Development Process Increases." Pew Research Center (Dec. 3, 2020). Accessed online at file:///C:/Users/Owner/AppData/Local/Temp/PS_ 2020.12.03_covid19-vaccine-intent_REPORT.pdf.

Funke, Daniel. "COVID-19 Vaccine Does Not Cause Death, Autoimmune Diseases." Politifact (The Poynter Institute) (Mar. 4, 2021). Accessed online at https://www.politifact.com/factchecks/2021/mar/04/sherri-tenpenny/covid-19-vaccine-does-not-cause-death-autoimmune-d/.

———. "Fact Check: COVID-19 Vaccines Don't Cause Death, Won't Decimate World's Population." *USA Today* (Apr. 30, 2021). Accessed online at https://www.usatoday.com/story/news/factcheck/2021/04/30/fact-check-covid-19-vaccines-dont-cause-death-wont-depopulate-planet/7411271002/.

Gavi (The Vaccine Alliance). "The COVID-19 Vaccine Race—Weekly Update." *Gavi* (Aug. 25, 2021). Accessed online 8/29/2021 at https://www.gavi.org/vaccineswork/covid-19-vaccine-race.

Gazit, Sivan, et al. "Comparing SARS-CoV-2 Natural Immunity to Vaccine-induced Immunity: Reinfections Versus Breakthrough Infections." medRxiv (2021). Accessed online at https://www.medrxiv.org/content/10.1101/2021.08.24.21262415v1.full.pdf.

Gherardi, Romain K., et al. "Biopersistence and Brain Translocation of Aluminum Adjuvants of Vaccines." *Frontiers in Neurology* 6, (2015) article 4.

Gherardi, Romain K., and Francois-Jerome Authier. "Macrophagic Myofasciitis: Characterization and Pathophysiology." *Lupus* 21/2 (2012) 184–89.

Gov.UK. "Deaths in Untied Kingdom." Gov.UK/Coronavirus (COVID-19) in the UK (Sept. 7, 2021). Accessed online at https://coronavirus.data.gov.uk/details/deaths.

Goyal, Rishi, Arden Hegele, and Dennis Tenen. "Op-Ed: How 'My Body, My Choice' Came to Define the Vaccine Skepticism Movement." *Los Angeles Times* (May 22, 2021). Accessed online at https://www.latimes.com/opinion/story/2021-05-22/vaccine-hesitancy-language-covid.

Haase, N., P. Schmid, and C. Betsch. "Impact of Disease Risk on the Narrative Bias in Vaccination Risk Perceptions." *Psychology and Health* 35/3 (2020) 346–65.

Haelle, Tara. "This Is the Moment the Anti-Vaccine Movement Has Been Waiting For." *The New York Times* Guest Essay (Aug. 31, 2021). Accessed online at https://www.nytimes.com/2021/08/31/opinion/anti-vaccine-movement.html.

Hahn, Ulrike, and Adam J. L. Harris. "What Does It Mean to Be Biased: Motivated Reasoning and Rationality." In *The Psychology of Learning and Motivation*, edited by B. H. Ross, 41–102. New York: Elsevier, 2014.

Harrington, Brooke. "The Anti-Vaccine Con Job Is Becoming Untenable." *The Atlantic* (Aug. 1, 2021). Accessed online at https://www.theatlantic.com/ideas/archive/2021/08/vaccine-refusers-dont-want-blue-americas-respect/619627/.

Hasöksüz, Mustafa, Selcuk Kiliç, and Fahriye Saraç. "Coronaviruses and SARS-COV-2." *Turkish Journal of Medical Sciences* 50 (2020) 549–56.

Hatfill, Steven J. "The Intentional Destruction of the National Pandemic Plan." *Journal of American Physicians and Surgeons* 26/3 (Fall, 2021) 74–76. Accessed online at https://www.jpands.org/vol26no3/hatfill.pdf.

Heinz, Franz X., and Karen Stiasny. "Distinguishing Features of Current COVID-19 Vaccines: Knowns and Unknowns of Antigen Presentation and Modes of Action." *NPJ Vaccines* (2021) 104. Accessed online at https://www.nature.com/articles/s41541-021-00369-6.pdf.

Held, Kristin S. "COVID-19 Statistics and Facts: Meaningful or a Means of Manipulation?" *Journal of American Physicians and Surgeons* 25/3 (Fall, 2020) 70–72. Accessed online at https://www.jpands.org/vol25no3/held.pdf.

Hinton, Denise M. "Letter to Ashley Rhoades." (Oct. 22, 2020). Accessed online at https://www.fda.gov/media/137564/download?mod=article_inline&mod=article_inline.

Hippisley-Cox, Julia. "Risk of Thrombocytopenia and Thromboembolism after Covid-19 Vaccination and SARS-CoV-2 Positive Testing: Self-controlled Case Series Study." *British Medical Journal* (2021) 374:n.1931.

Hollingsworth, Heather, Cathy Bussewitz, and Colleen Long. "COVID-19 Cases Climbing, Wiping Out Months of Progress." *AP News* (Sept.14, 2021). Accessed online at https://apnews.com/article/joe-biden-health-texas-coronavirus-pandemic-kentucky-9e89fcef6ef184baa4703a47d2aefab6.

Hooker, Brian S., and Neil Z. Miller. "Analysis of Health Outcomes in Vaccinated and Unvaccinated Children: Developmental Delays, Asthma, Ear Infections, and Gastrointestinal Disorders." *SAGE Open Medicine* 8 (2020) 1–11.

Hotez, Peter J., et al. "COVID-19 Vaccines: Neutralizing Antibodies and the Alum Advantage." *Nature Reviews* 20 (July, 2020) 399–400.

Hu, Ben, et al. "Reviews: Characteristics of SARS-CoV-2 and COVID-19." *Nature Reviews Microbiology* 19 (Mar., 2021) 141–54.

Iacobucci, Gareth. "Long Covid: Damage to Multiple Organs Presents in Young, Low Risk Patients." *The BMJ* 371 (17 Nov., 2020) 4470. Accessed online at https://www.bmj.com/content/bmj/371/bmj.m4470.full.pdf.

Ipsos. "The Wall of Vaccine Opposition Might Be Starting to Crumble." (Aug. 31, 2021). Accessed online at https://www. ipsos.com/en-us/news-polls/axios-ipsos-coronavirus-index.

Ipsos/Axios Poll (March 19-22) Accessed online at https://www.ipsos.com/sites/default/files/ct/news/documents/2021-03/topline-axios-ipsos-coronavirus-index-w42.pdf.

Iwasaki, Akiko, and Yexin Yang. "The Potential Danger of Suboptimal Antibody Responses in COVID-19." *Nature Reviews Immunology* 20 (2020) 339–41. Accessed online at https://www.nature.com/articles/s41577-020-0321-6.

Jarry, Jonathan. "The Anti-Vaccine Movement in 2020." McGill: Office for Science and Society (May 22, 2020). Accessed online at https://www.mcgill.ca/oss/article/covid-19-pseudoscience/ anti-vaccine-movement-2020.

———. "The Physician Who Calmly Denies Reality." McGill: Office for Science and Society (Sept. 24, 2020). Accessed online at https://www.mcgill.ca/oss/article/covid-19-pseudoscience/psychiatrist-who-calmly-denies-reality.

Johnson, N.F., et al. "The Online Competition between Pro- and Anti-vaccination Views." *Nature* 582 (2020) 230–33.

Joint Statement in Support of COVID-19 Vaccine Mandates for All Workers in Health and Long-Term Care (July 26, 2021). Accessed online at https://leadingage.org/sites/default/files/Joint%20Statement%20on %20Vaccine%20Mandates.pdf.

Jolley, Daniel, and Karen M. Douglas. "Prevention Is Better than Cure: Addressing Anti-vaccine Conspiracy Theories." *Journal of Applied Social Psychology* 47/8 (2017) 459–69.

Kahn, Jo. "We've Never Made a Successful Vaccine for a Coronavirus Before. This Is Why It's So Difficult." ABC News (Australia) (Apr. 16, 2020). Accessed online at https://www.abc.net.au/news/health/2020-04-17/coronavirus-vaccine-ian-frazer/12146616.

Kanefield, Teri. "What Could Be Motivating the Wild Covid Recklessness in States Like Florida." NBC News (Sept. 25, 2021). Accessed online at https://www.msn.com/en-us/news/politics/what-could-be-motivating-the-wild-covid-recklessness-in-states-like-florida/ar-AAONLBO?ocid=msedgdhp&pc=U531.

Kayyim, Juliette. "Vaccine Refusers Don't Get to Dictate Terms Anymore." *The Atlantic* (Aug. 29, 2021). Accessed online 8/29/2021 at https://www.msn.com/en-us/news/us/vaccine-refusers-don-t-get-to-dictate-terms-anymore/ar-AANRIvX?ocid=msedgdhp&pc=U531.

Kempen, Paul Martin. "A Perspective on Our Time with COVID." *Journal of American Physicians and Surgeons* 25/4 (Winter, 2020) 109–11. Accessed online at https://www.jpands.org/vol25no4/kempen.pdf.

Kennedy, Robert F., Jr. "Standing Up for Our Children." Children's Health Defense. Accessed online at https://childrens healthdefense.org/ebook-sign-up-vaccine-mandates-an-erosion-of-civil-rights/.

Khazan, Olga. "The Opposite of Socialized Medicine." *The Atlantic* (Feb. 25, 2020). Accessed online at https://www.theatlantic .com/health/archive/2020/02/aaps-make-health-care-great-again/607015/.

Kim, Dongwan, et al. "The Architecture of SARS-CoV-2 Transcriptome." *Cell* 181 (May 14, 2020) 914–21.

Kortum P, C. Edwards, and R. Richards-Kortum. "The Impact of Inaccurate Internet Health Information in a Secondary School Learning Environment." *Journal of Medical Internet Research* 10/2 (2008) e17.

Lam-Hine, Tracy, et al. "Outbreak Associated with SARS-CoV-2 B.1.617.2 (Delta) Variant in an Elementary School — Marin County, California, May–June 2021." CDC *Morbidity and Mortality Weekly Report (MMWR)* 70 (Aug. 27, 2021) 7035e2. Accessed online at https://www.cdc.gov/mmwr/volumes/70/wr/mm7035 e2.htm?s_cid=mm7035e2_w.

Lapado, Joseph A. "An American Epidemic of 'Covid Mania.'" *Wall Street Journal* Opinion/Commentary (Apr. 19, 2021). Accessed online at https://www.wsj.com/articles/an-american-epidemic-of-covid-mania-11618871457.

Ledford, Heidi. "Six Months of COVID Vaccines: What 1.7 Billion Doses Have Taught Scientists." *Nature* 594 (2021) 164–67. Accessed online at https://www.nature .com/articles/d41586-021-01505-x.

Le Guillou, Ian. "Covid-19: How Unprecedented Data Sharing Has Led to Faster-than-Ever Outbreak Research." *Horizon: The EU Research and Innovation Magazine* (Mar. 23, 2020). Accessed online at https://ec.europa.eu/research-and-innovation/en/horizon-magazine/covid-19-how-unprecedented-data-sharing-has-led-faster-ever-outbreak-research.

Lehrer, Steven, and Peter H. Rheinstein. "Human Gene Sequences in SARS-CoV-2 and Other Viruses." *In Vivo* 34 (2020) 1633–36. Accessed online at https://iv.iiarjournals.org/content/invivo/34/3_suppl/1633.full.pdf.

Lei, Yuyang, et al. "SARS-CoV-2 Spike Protein Impairs Endothelial Function via Downregulation of ACE 2." *Circulation Research* 128/9 (Apr. 30, 2021) 1323–26. Accessed online at https://www.ahajournals.org/doi/epub/10.1161/CIRCRESAHA.121.318902.

"Let's Talk about Lipid Nanoparticles." *Nature Reviews Materials* 6 (Feb., 2021) 99. Accessed online at https://www.nature.com/articles/s41578-021-00281-4.pdf.

Leung, Angela Ki Che. "'Variolation' and Vaccination in Late Imperial Chine, Ca 1570–1911." In *History of Vaccine Development*, edited by Stanley A. Plotkin, 5–12. New York: Springer, 2011.

Lewis, Jarrett Ramos. "What Is Driving the Decline in People's Willingness to Take the COVID-19 Vaccine in the United States?" *JAMA Health Forum* 1/11 (2020) e201393.

Li, Jianyu and David J. Mooney. "Designing Hydrogels for Controlled Drug Delivery." *Nature Reviews Materials* 1/12 (2016) 16071. Accessed online at https://www.ncbi.nlm.nih.gov/pmc/articles/PMC5898614/pdf/nihms955923.pdf.

Liang, Zhihui, et al. "Adjuvants for Coronavirus Vaccines." *Frontiers in Immunology* 11 (Nov., 2020) Article 589833.

Liu, Angus. "An mRNA Vaccine Delivered in Hydrogel Shows Promise as a Durable Cancer Immunotherapy." Fierce BioTech (Feb. 17, 2021). Accessed online at https://www.fiercebiotech.com/research/mrna-vaccine-delivered-hydrogel-shows-promise-as-a-durable-cancer-immunotherapy.

Liu, Li, et al. "Anti-Spike IgG Causes Severe Acute Lung Injury by Skewing Macrophage Responses During Acute SARS-CoV Infection." *JCI Insight* 4/4 (2019) e123158. Accessed online at https://web.archive.org/web/20210228134922/https://insight.jci.org/articles/view/123158/pdf.

Lock, Samantha. "Dangerously Mutated R.1 COVID Variant Detected in 35 Countries." *Newsweek* (Sept. 23, 2021). Accessed online at https://www.msn.com/en-us/health/medical/dangerously-mutated-r-1-covid-variant-detected-in-35-countries/ar-AAOKegI?ocid=msedgdhp&pc=U531.

Lombard, M., P.-P. Pastoret, and A.-M. Moulin. "A Brief History of Vaccines and Vaccination." *Scientific and Technical Review of the Office International des Epizooties* 26/1 (2007) 29–48.

Lovett, Ian. "White Evangelicals Resist COVID-19 Vaccine Most among Religious Groups." *Wall Street Journal* (July 28, 2021). Accessed online at https://www.wsj.com/articles/white-evangelicals-resist-covid-19-vaccine-most-among-religious-groups-11627464601.

Lyons-Weller, James, and Paul Thomas. "Relative Incidence of Office Visits and Cumulative Rates of Billed Diagnoses along the Axis of Vaccination." *International Journal of Environmental Research and Public Health* (Nov. 22, 2020). Accessed online at https://www.mdpi.com/1660-4601/17/22/8674.

Machingaidze, Shingai, and Charles Shey Wiysonge. "Understanding COVID-19 Vaccine Hesitancy." *Nature Medicine* 27 (2021) 1338–39.

Manderson, Nathaniel. "Evangelicals, Science and the Vaccine: Refusal Is Built on Deep-seated Fear." *Salon* (Aug. 28, 2021). https://www.msn.com/en-us/news/opinion/ evangelicals-science-and-the-vaccine-refusal-is-built-on-deep-seated-fear/ar-AANQHjL?ocid=msedgdhp&pc=U531.

Maragakis, Lisa Lockerd [reviewer]. "Covid-19 vs. the Flu." Johns Hopkins Medicine (July 29, 2021 update). Accessed online at https://www.hopkinsmedicine.org/health/conditions-and-diseases/coronavirus/coronavirus-disease-2019-vs-the-flu.

Marano, Marc. "COVID-19 as a Model for the Climate Change Agenda." *Journal of American Physicians and Surgeons* 26/3 (Fall, 2021) 77–82. Accessed online at https://www.jpands.org/vol26no3/morano.pdf.

Martiny, Clara, and Kayla Gogarty. "Facebook Groups around the World Are Promoting Unprescribed Livestock Medications for COVID-19, While the Platform Seemingly Does Nothing to Stop Them." Media Matters for America (Updated Sept. 1, 2021). Accessed online at https://www.mediamatters.org/facebook/facebook-groups-around-world-are-promoting-unprescribed-livestock-medications-covid-19.

Matheson, Nicholas J., and Paul J. Lehner. "How Does SARS-CoV-2 Cause COVID-19?" *Science* 369 (July 31, 2020) Issue 6593, 510–11.

Matthews, David. "Idaho Nurse Who Refused COVID Vaccination Dies: Brother." NY Daily News.com. (Sept. 21, 2021). Accessed online at https://www.msn.com/en-us/news/us/idaho-nurse-who-refused-covid-vaccine-dies-brother/ar-AAOFHNh?ocid=msedgdhp&pc=U531.

McBean, Eleanor. *The Poisoned Needle: Suppressed Facts about Vaccination*. Mokelumne Hill, CA: Health Research, 1956.

McDonald, Jessica. "Posts Misinterpret CDC's Provincetown COVID-19 Outbreak Report." FactCheck.org (Aug. 6, 2021). Accessed online at https://www.factcheck.org/2021/08/scicheck-posts-misinterpret-cdcs-provincetown-covid-19-outbreak-report/.

McDonnell, Anthony, and Flavio Toxvaerd. "How Does the Market for Vaccines Work?" *Economics Observatory* (May 13, 2021). Accessed online at https://www.economicsobservatory.com/how-does-the-market-for-vaccines-work.

Melendez, Pilar. "Demon Sperm's Doc's Pals Launch Twisted New Crusade to Stop Vaccines." *Daily Beast* (July 20, 2021). Accessed online at https://www.thedailybeast.com/the-most-dangerous-and-deranged-claims-in-americas-frontline-doctors-motion-against-covid-vaccinations.

Mercer, Mia. "Why Evangelicals Are Encouraging the Anti-Vaccination Movement. Liberal Arts, Texas A&M Univ. (May 4, 2021). Accessed online at https://liberalarts.tamu.edu/blog/2021/05/04/why-evangelicals-are-encouraging-the-anti-vaccination-movement/.

Miller, Janel. "Remdesivir, Hydroxychloroquine Fail to Show Antiviral Effects in Patients with COVID-19." Healio (July 12, 2021). Accessed online at https://www.healio.com/news/primary-care/20210712/remdesivir-hydroxychloroquine-fail-to-show-antiviral-effects-in-patients-with-covid19.

Millet, Jean Kaoru, and Gary R. Whittaker. "Host Cell Proteases: Critical Determinants of Coronavirus Tropism and Pathogenesis." *Virus Research* 202 (2015) 120–34.

Miranda, Gabriela. "You're More Likely to Get a Blood Clot after COVID Infection than with the Vaccine." *USA Today* (Aug. 27, 2021), accessed online at https://www.msn.com/en-us/health/medical/you-re-more-likely-to-get-a-blood-clot-after-covid-infection-than-with-the-vaccine/arAANOAPB?ocid=msedgdhp&pc=U531.

Moran, Meghan Bridgid. "Anti-Vaxx Websites, We're Unto You." *Time* (Feb. 11, 2016). Accessed online at https://time.com/4213054/anti-vaxx-websites/.

Moran, Meghan Bridgid, et al. "Why Are Anti-Vaccine Messages So Perusasive? A Content Analysis Anti-Vaccine Websites' Techniques to Engender Anti-Vaccine Sentiment." Academia. Accessed online at https://www.academia.edu/20554328/Why_are_anti-vaccine_messages_so_persuasive_A_content_analysis_anti-vaccine_websites_techniques_to_engender_anti-vaccine_sentiment.

Mower, Lawrence, and Kirby Wilson. "Florida's Next Surgeon General Opposes Mask, Vaccine Mandates." *Miami Herald* (Updated Sept. 21, 2021). Accessed online at https://www.miamiherald.com/news/coronavirus/article254412224.html.

National Vaccine Information Center (NVIC). "SARS CoV-2 Virus and COVID-19 Vaccine Information." NVIC (*Updated July 22, 2021*). Accessed online at https://www.nvic.org/vaccines-and-diseases/covid-19.aspx#_edn6.

"Natural Immunity vs. Artificial Immunity." Dangers of Vaccines. Accessed online at http://dangersofvaccines.com/history-of-vaccines/natural-immunity-vs-artificial-immunity/.

NIH (National Institutes of Health). "Ivermectin." (Updated Feb. 11, 2021). Accessed online at https://www.covid19treatment guidelines.nih.gov/therapies/antiviral-therapy/ivermectin/.

———. "Remdesivir." (Updated Apr. 21, 2021). Accessed online at https://www.covid19treatmentguidelines.nih.gov/therapies/antiviral-therapy/remdesivir/.

Nishiura, Hiroshi, Natalie M. Linton, and Andrei R. Akhmetzhanov. "Serial Interval of Novel Coronavirus (COVID-19) Infections." *International Journal of Infectious Diseases* 93 (2020) 284–86.

Noble, Alex. "Caleb Wallace, Anti-Mask 'Freedom Rally' Organizer, Dies at 30 with COVID." *The Wrap* (Aug. 28, 2021). Accessed online at https://www.msn.com/en-us/tv/news/caleb-wallace -anti-mask-freedom-rally-organizer-dies-at-30-with-covid/ar-AANQY7H?ocid=msedgdhp&pc=U531.

Offit, Paul A. "Why Are Pharmaceutical Companies Gradually Abandoning Vaccines?" *Health Affairs* 24/3 (2005) 622–30.

Offit, Paul A., et al. "Addressing Parents' Concerns: Do Multiple Vaccines Overwhelm or Weaken the Infant's Immune System?" *Pediatrics* 109 (2002) 124–29.

Orient, Jane M. "COVID vs. the Oath of Hippocrates." *Journal of American Physicians and Surgeons* 25/4 (Winter, 2020) 105–08. Accessed online at https://www.jpands.org/vol25no4/orient.pdf.

———. "Statement on Federal Vaccine Mandates." (Feb. 26, 2019). Accessed online at https://aapsonline.org/measles-outbreak-and-federal-vaccine-mandates/.

———. "Testimony before the U.S. Senate Committee on Homeland Security and Government Oversight (December 8, 2020)." Accessed online at https://www.hsgac.senate.gov/imo/media/doc/Testimony-Orient-2020-12-08.pdf

Pardi, Norbert, et al. "mRNA Vaccines—A New Era in Vaccinology." *Nature Reviews Drug Discovery* 17 (Apr., 2018) 261–79. Accessed online at https://www.nature.com/articles/nrd.2017.243.pdf.

Parsons, Rachael. "Faith, Trust or Science?—The COVID Vaccine, Part 1." Vaxxter. (Jan. 18, 2021). Accessed online at https://vaxxter.com/faith-trust-or-science-the-covid-vaccine-part-1/.

"Pediatrician's License Suspended in Oregon over Vaccines." *Modern Health Care* (Dec. 9, 2020). Accessed online at https://www.modernhealthcare.com/physicians/pediatricians-license-suspended-oregon-over-vaccines.

Petousis-Harris, Helen. "Assessing the Safety of COVID-19 Vaccines: A Primer." *Drug Safety* 43 (2020) 1205–10. Accessed online at https://link.springer.com/content/pdf/10.1007/s40264-020-01002-6.pdf.

Pew Research Center. "Intent to Get Vaccinated Varies by Religious Affiliation in the U.S." (Mar. 22, 2021). Accessed online at https://www.pewresearch.org/fact-tank/2021/03/23/10-facts-about-americans-and-coronavirus-vaccines/ft_21-03-18 _ vaccinefacts/.

Pies, Ronald W. "Anti-Vaxxers and Water Witches: Mistrust of Science and the Limits of Reason." *Psychiatric Times* (July 22, 2021). Accessed online at https://www.psychiatrictimes.com/view/anti-vaxxers-water-witches-mistrust-of-science-limits-of-reason.

Porter, Tom. "How the Evangelical Christian Right Seeded the False, Yet Surprisingly Resilient, Theory that Vaccines Contain Microchips." Business Insider (Sept. 24, 2021). Accessed online at https://www.businessinsider.com/how-evangelical-right-pushed-microchip-vaccine-conspiracy-theory-2021-9.

Price, Robert. "Controversial Bakersfield Doctor Back in National Spotlight after Aligning with Group Downplaying COVID-19 Threat." KGET.com (July 30, 2020). Accessed online at https://www.kget.com/health/coronavirus/controversial-bakersfield-doctor-back-in-national-spotlight-after-aligning-with-group-downplaying-covid-19-threat/.

Rainie, Lee, and Andrew Perrin. "The State of Americans' Trust in Each Other Amid the COVID-19 Pandemic." Pew Research Center (Apr. 6, 2020). Accessed online at https://www. pewresearch.org/fact-tank/2020/04/06/the-state-of-americans-trust-in-each-other-amid-the-covid-19-pandemic/.

Randolph, Haley E., and Luis B. Barreiro. "Herd Immunity: Understanding COVID-19." *Immunity* 52 (May 19, 2020). 737–41. Accessed online at https://reader.elsevier.com/reader/sd/pii/S1074761320301709?token=1ED532D41F118266FB1073455F939011F2629303FACE205F4186D78BB6312F8AB39EE99239FC9F5D4ECFC8885061DCAB&originRegion=us-east1&originCreation=2021090921 1624.

Rawls, John. *A Theory of Justice*. Cambridge: Harvard Univ. Press, 1999. (Original work published 1971.)

Reardon, Sara. "Flawed Ivermectin Preprint Highlights Challenges of COVID Drug Studies." *Nature* 596 (2021) 173–74.

Reiss, Karina, and Sucharit Bhakdi. *Corona False Alarm? Facts and Figures*. White River Junction: Chelsea Green Publishing, 2020.

"Retraction. Ileal-lymphoid-nodular hyperplasia, non-specific colitis, and pervasive developmental disorder." *Lancet* 375/9713 (2010) 445.

Reuben, Rebekas, Devon Aitken, Jonathan L. Freedman, and Gillian Einstein. "Mistrust of the Medical Profession and Higher Disgust Sensitivity Predict Parental Vaccine Hesitancy." *PloS One* (Sept. 2, 2020). Accessed online at https://www.ncbi.nlm.nih.gov/pmc/articles/PMC7467323/pdf/pone.0237755.pdf.

Reuters Fact Check. "Fact Check-mRNA Vaccines Do Not Turn Humans into 'Hybrids' or Alter Recipients' DNA." Reuters (Apr. 13, 2021). Accessed online at https://www.reuters.com/ article/factcheck-mrna-megamix/fact-check-mrna-vaccines-do-not-turn-humans-into-hybrids-or-alter-recipients-dna-idUSL1N2M61HW.

Reuters Staff. "Fact Check: Available mRNA Vaccines Do Not Target Syncitin-1, a Protein Vital to Successful Pregnancies." Accessed online at https://www.reuters.com/ article/uk-factcheck-syncytin/fact-check-available-mrna-vaccines-do-not-target-syncytin-1-a-protein-vital-to-successful-pregnancies-idUSKBN2A42S7.

———. "Fact Check: COVID-19 Vaccines Do Not Contain the Ingredients Listed in These Posts." (Feb. 26, 2021). Accessed online at https://www.reuters.com/article/uk-factcheck-covid-vaccine-ingredients/fact-check-covid-19-vaccines-do-not-contain-the-ingredients-listed-in-these-posts-idUSKBN2 AQ2SW.

———. "False Claim: A COVID-19 Vaccine Will Genetically Modify Humans." Reuters (May 18, 2020). Accessed online at https://www.reuters.com/article/uk-factcheck-covid-19-vaccine-modify/false-claim-a-covid-19-vaccine-will-genetically-modify-humans-idUSKBN22U2BZ.

Rice, Nicholas. "Conservative Radio Host Who Called Himself 'Mr. Anti-Vax' Dies from COVID After 3 Week Battle." *People* (Aug. 30, 2021). Accessed online at https://www.msn.com/en-us/tv/news/conservative-radio-host-who-called-him

self-mr-anti-vax-dies-from-covid-after-3-week-battle/ar-AANUuqr?ocid=msedgdhp&pc=U531.

Romo, Vanessa. "Poison Control Centers Are Fielding a Surge of Ivermectin Overdose Calls." NPR (National Public Radio) (Sept. 4, 2021). Accessed online at https://www.npr.org/sections/coronavirus-live-updates/2021/09/04/1034217306/ivermectin-overdose-exposure-cases-poison-control-centers.

Rothwell, Jonathan, and Dan Witters. "U.S. Adults' Estimates of COVID-19 Hospitalization Risk." Gallup (Sept. 27, 2021). Accessed online at https://news.gallup.com/opinion/gallup/354938/adults-estimates-covid-hospitalization-risk.aspx.

Sajjadi, Nicholas B., et al. "United States Internet Searches for 'Infertility' Following COVID-19 Vaccine Misinformation." *Journal of Osteopathic Medicine* 121/6 (2021) 583–87.

Salamanna, Francesca, et al. "Body Localization of ACE-2: On the Trail of the Keyhole of SARS-CoV-2." *Frontiers in Medicine* 7 (Dec., 2020) Art. 594495. Accessed online at https://www.frontiersin.org/articles/10.3389/fmed.2020.594495/full.

Sanche, Steven, et al. "The Novel Coronavirus, 2019-nCoV, Is Highly Contagious and More Infectious than Initially Estimated." medRxiv preprint (Feb. 11, 2020). Accessed online at https://www.medrxiv.org/content/10.1101/2020.02.07.20021154v1.full.pdf.

Santucci, Jeanine. "Unvaccinated Are 11 Times More Likely to Die of COVID-19, CDC Studies Show: Latest Updates." *USA Today*. (Sept. 27, 2021). Accessed online at https://www.usatoday.com/story/news/health/2021/09/11/cdc-unvaccinated-study-covid-live-updates/8277456002/.

Sanyal, Sumana. "How SARS-CoV-2 (COVID-19) Spreads Within Infected Hosts—What We Know So Far." *Emerging Topics in Life Sciences* 4 (2020) 383–90.

Schimelpfening, Nancy. "No, Fetal Tissues Weren't Used to Create the J&J COVID-19 Vaccine." Healthline (Mar. 18, 2021). Accessed online at https://www.healthline.com/health-news/no-fetal-tissue-wasnt-used-to-create-the-jj-covid-19-vaccine.

Schoeman, Dewald, and Burtram C. Fielding. "Coronavirus Envelope Protein: Current Knowledge." *Virology Journal* 16 (2019) 69. Accessed online at https://virologyj.biomedcentral.com/track/pdf/10.1186/s12985-019-1182-0.pdf.

Science Media Centre. "Expert Reaction to Study Looking at Risk of Thrombocytopenia and Thromboembolism after COVID-19 Vaccination and SARS-CoV-2 Infection." Accessed online at https://www.sciencemediacentre.org/expert-reaction-to-study-looking-at-risk-of-thrombocytopenia-and-thromboembolism-after-covid-19-vaccination-and-sars-cov-2-infection/.

Seidel, Kathleen. "Strange Bedfellows." *Neurodiversity weblog* (Mar. 12, 2006). Accessed online at https://web.archive.org/web/20110927235939/http://www.neurodiversity.com/weblog/article/91/.

Shang, Jian, et al. "Cell Entry Mechanisms of SARS-CoV-2." *Proceedings of the National Academy of Sciences (PNAS)* 117/21 (May 26, 2020) 11727–34.

Singleton, Marilyn M. "COVID-19: A Weapon to Fundamentally Transform America." *Journal of American Physicians and Surgeons* 26/2 (Summer, 2021) 43–50. Accessed online at https://www.jpands.org/vol26no2/singleton.pdf.

Sinha, Neeraj, and Galit Balayla. "Hydroxychloroquine and COVID-19." *Postgraduate Medical Journal* 96 (2020) 550–55.

Smith, Rory, Liliana Bounegru, and Jonathan Gray. "How Anti-vaccination Websites Build Audiences and Monetize Misinformation." First Draft (Mar. 10, 2021). Accessed online at https://firstdraftnews.org/articles/antivaccination-audiences-monetize/.

Sparks, Grace, Ashley Kirzinger, and Mollyann Brodie. "KFF COVID-19 Vaccine Monitor: Profile of the Unvaccinated." Kaiser Family Foundation (KFF) (June 11, 2021). Accessed online at https://www.kff.org/coronavirus-covid-19/poll-finding/kff-covid-19-vaccine-monitor-profile-of-the-unvaccinated/.

Stolberg, Sheryl Gay. "Anti-Vaccine Doctor Has Been Invited to testify Before Senate Committee." *The New York Times* (Dec. 6, 2020; updated Apr. 26, 2021). Accessed online at https://www.nytimes.com/2020/12/06/us/politics/anti-vax-scientist-senate-hearing.html.

Stolle, Lucas B., et al. "Fact vs Fallacy: The Anti-Vaccine Discussion Reloaded." *Advances in Therapy* 37 (2020) 4481–90.

Strasburg, Jenny, and Cecilia Butini, "AstraZeneca Loses Money on Covid-19 Vaccine for Second Straight Quarter." *Wall Street Journal* (July 29, 2021). Accessed online at https:// www.wsj.com/articles/ astrazeneca-loses-money-on-covid-19-vaccine-for-second-straight-quarter-11627547543.

Statista. "Average Number of Annual Smallpox Deaths Per Million Inhabitants in England during the Various Stages of Vaccination Implementation between 1700 and 1898." Accessed online at https://www.statista.com/statistics/1107661/smallpox-vaccination-impact-england-historical/.

Stecklow, Steve, and Andrew Macaskill "The Ex-Pfizer Scientist Who Became an Anti-Vax Hero." Reuter's Special Report (Mar. 18, 2021). Accessed online at https://www.reuters.com/investigates/special-report/health-coronavirus-vaccines-skeptic/.

Supreme Court of the United States. "Syllabus: Bruesewitz et al. v. Wyeth et al." (2011). Accessed online at https://www.supremecourt.gov/opinions/10pdf/09-152.pdf.

Swenson, Ali. "US and EU COVID Vaccines Don't Contain Alunimum." AP News (Mar. 16, 2021). Accessed online at https://apnews.com/article/fact-checking-afs:Content:99910 20426.

Taquest, M. et al. "Cerebral Venous Thrombosis and Portal Vein Thrombosis: A Retrospective Cohort Study of 537913 COVID-19 cases. *Open Science Framework (OSF)* (2021).

Tenpenny, Sherr. "Coronavirus Pt. 6: The COVID Vaccines—Part 2—UPDATED." Vaxxter. (Dec. 28, 2020). Accessed online at https://web.archive.org/web/20210228231513/https://vaxxter.com/covid-vaccines-part-2/.

Turner, Jackson S., et al. "SARS-CoV-2 Infection Induces Long-lived Bone Marrow Plasma Cells in Humans." *Nature* 595 (July 15, 2021) 421–25. Accessed online at https://www.nature.com/articles/s41586-021-03647-4.pdf.

Twohig, Katherine A., et al. "Hospital Admission and Emergency Care Attendance Risk for SARS-CoV-2 delta (B.1.617.2) Compared with Alpha (B.1.17) Variants of Concern: A Cohort Study." *The Lancet* (2021). Published onlie Aug. 27, 2021. Accessed online at https://www.thelancet.com/action/ showPdf?pii=S1473-3099%2821%2900475-8.

V'kovski, Philip, et al. "Coronavirus Biology and Replication: Implications for SARS-CoV-2." *Nature Reviews Microbiology* 19 (Mar., 2021) 155–70.

VAERS (Vaccine Adverse Event Reporting System). "Disclaimer." Accessed online at https://vaers.hhs.gov/data.html.

Vallentyne, Peter (2008). "Brute luck and responsibility," *Politics, Philosophy & Economics* 7/1 (2008) 57–80.

Vaughan, Brandy. "Do You Know What's in a Vaccine?" Learn the Risk (2018). Accessed online at https://learntherisk.org/vaccines/ingredients/.

Vaxxter. "New Study Confirms: The COVID Shot Spike Protein Is Dangerous." Vaxxter (May 4, 2021). Accessed online at https://web.archive.org/web/20210916192807/https://vaxxter.com/new-study-confirms-the-covid-shot-spike-protein-is-dangerous/.

———. "New Survey of Vaccine-Free Group Exposes Long-term Impact of Vaccination Policies on Public Health by Greg Glaser and Pat O'Connell." Vaxxter (Nov. 30, 2020). Accessed online at https://web.archive.org/web/20210916192813/https://vaxxter.com/new-survey-of-vaccine-free-group-exposes-long-term-impact-of-vaccination-policies-on-public-health-by-greg-glaser-and-pat-oconnell/.

Velan, Baruch, et al. "Individualism, Acceptance and Differentiation as Attitude Traits in the Public's Response to Vaccination." *Human Vaccines and Immunotherapeutics* 8/9 (Sept., 2012) 1272–82. Accessed online at https://www.ncbi.nlm.nih.gov/pmc/articles/PMC3579908/pdf/hvi-8-1272.pdf.

Wadman, Meredith. "Abortion Opponents Oppose COVID-19 Vaccines' Use of Fetal Cells." *Science*/News (June 5, 2020). Accessed online at https://www.science.org/news/2020/06/abortion-opponents-protest-covid-19-vaccines-use-fetal-cells.

Wakefield, Andrew, et al. "Ileal-lymphoid-nodular hyperplasia, non-specific colitis, and pervasive developmental disorder." *Lancet* 351/9103 (1998) 637–41.

Wang, Manli, et al. "Remdesivir and Chloroquine Effectively Inhibit the Recently Emerged Novel Coronavirus (2019-nCoV) in Vitro." *Cell Research* 30 (2020) 269–41. Accessed online at https://www.ncbi.nlm.nih.gov/pmc/articles/PMC7054408/pdf/41422_2020_Article_282.pdf.

Wax, Craig M. "The Critical Role of Early Home Treatment in Surviving and Thriving after COVID-19." *Journal of American Physicians and Surgeons* 26/1 (Spring, 2021) 19–20. Accessed online at https://www.jpands.org/vol26no1/wax.pdf.

Weintraub, Arlene. "Pfizer CEO Says It's 'Radical' to Suggest Pharma Should Forgo Profits on COVID-19 Vaccine: Report." *Fierce Pharma* (July 30, 2020). Accessed online at https://www.fiercepharma.com/pharma/pfizer-ceo-says-it-s-radical-to-suggest-pharma-should-forgo-profits-covid-19-vaccine-report.

Weir, Keziah. "How Robert F. Kennedy Jr. Became the Anti-Vaxxer Icon of America's Nightmares." *Vanity Fair* (May 13, 2021). Accessed online at https://www.vanityfair.com/news/2021/05/how-robert-f-kennedy-jr-became-anti-vaxxer-icon-nightmare.

Welsh, Nick. "Santa Barbara Coroner Concludes Anti-Vaxxer Brandy Vaughan Died of Natural Causes." *Santa Barbara Independent* (Feb. 18, 2021). Accessed online at https://www.independent.com/2021/02/18/santa-barbara-coroner-concludes-anti-vaxxer-brandy-vaughan-died-of-natural-causes/.

Wertheimer, Ellen. "Unavoidably Unsafe Products: A Modest Proposal." *Chicag-Kent Law Review* 72/1 (1996) 189–217.

Westbrock, David A. "COVID-19: Reflections on the Disease and the Response." *Journal of American Physicians and Surgeons* 25/2 (Summer, 2020) 54–55. Accessed online at https://www.jpands.org/vol25no2/westbrock.pdf.

WHO (World Health Organization). "WHO Advises that Ivermectin Only Be Used within Clinical Trials." WHO (Mar. 31, 2020). Accessed online at https://www.who.int/news-room/feature-stories/detail/who-advises-that-ivermectin-only-be-used-to-treat-covid-19-within-clinical-trials.

WHO Solidarity Trial Consortium. "Repurposed Anti-viral Drugs for COVID-19—Interim WHO SOLIDARITY Trial Results." *New England Journal of Medicine* 384 (2021) 497–511. Accessed online at https://www.nejm.org/doi/pdf/10.1056/NEJMoa2023184?articleTools=true.

Williams, Gareth. *Angel of Death: The Story of Smallpox*. New York: Palgrave McMillan, 2010.

World Health Organization (WHO). "Ten Threats to Global Health in 2019." Accessed online at https://www. who.int/news-room/spotlight/ten-threats-to-global-health-in-2019.

Wu, Qianhui, et al., "Evaluation of the Safety Profile of COVID-19 Vaccines: A Rapid Review." *BMC Medicine* 19 (2021) 173. Accessed online at https://bmcmedicine.biomedcentral.com/track/pdf/10.1186/s12916-021-02059-5.pdf.

Yang, Yang, and Lanying Du. "SARS-CoV-2 Spike Protein: A Key Target for Eliciting Persistent Neutralizing Antibodies." *Signal Transduction and Targeted Therapy* 6 (2021) 95. Accessed online at https://www.nature.com/articles/s41392-021-00523-5.pdf.

Yeadon, Michael. "The PCR False Positive Pseudo-Epidemic." *The Daily Skeptic* (Nov. 30, 2020/updated Dec. 25, 2020). Accessed online at https://dailysceptic.org/the-pcr-false-positive-pseudo-epidemic/.

Yeadon, Michael, and Wolfgang Wodarg. "Petition/Motion for Administrative/Regulative Action, *etc*." Accessed online at file:///C:/Users/Owner/Downloads/Wodarg_Yeadon_EMA_Petition_Pfizer_Trial_ FINAL_01DEC2020 _signed_with _Exhibits_geschwärzt.pdf.

Zimmerman, Richard K. "Helping Patients with Ethical Concerns about COVID-19 Vaccines in Light of Fetal Cell Lines Used in Some COVID-19 Vaccines." *Vaccine* 39 (2021) 4242–44. Accessed online at https://www.ncbi.nlm.nih.gov/pmc/articles/PMC8205255/pdf/main.pdf.

N.B. Web links must be reproduced exactly to work, and even then materials are subject to removal or relocation. So searching by author and/or title may sometimes be advisable.

www.ingramcontent.com/pod-product-compliance
Lightning Source LLC
Chambersburg PA
CBHW070436180526
45158CB00019B/1467